THE POLITICS OF
NUCLEAR
PROLIFERATION

THE POLITICS OF NUCLEAR PROLIFERATION

George Quester

THE JOHNS HOPKINS UNIVERSITY PRESS

Baltimore and London

Written under the auspices of the Harvard University Center
for International Affairs and the Cornell University Program
on Peace Studies.

The Johns Hopkins University Press, Baltimore, Maryland 21218
The Johns Hopkins University Press Ltd., London

Library of Congress Catalog Card Number 73-8119
ISBN 0-8018-1477-4

Library of Congress Cataloging in Publication data
will be found on the last printed page of this book.

FOR TEDDY AND AMY

Contents

Preface

The nuclear proliferation question is likely to be present in world politics for a long time. The possibility that many nations might learn to produce nuclear weapons became a reality in the 1960s, and the Nuclear Non-Proliferation Treaty (NPT) then emerged as a deliberate and calculated effort to head it off. Yet the NPT alone cannot eliminate the prospect of nuclear proliferation; at best it can put a freeze on it until fundamental changes appear in world politics which eliminate all possible national incentives for seeking such weapons. One may become prematurely reassured about nuclear proliferation, as well as prematurely alarmed, while "world politics as usual" continues all the while. The purpose of this study is specifically to examine the ways in which international politics has exploited the possibility of nuclear proliferation and the treaty intended to prevent it.

It would be difficult to undertake any study of nuclear proliferation without the assistance of a great number of people. I should first acknowledge the generosity of the many individuals, in and out of the various relevant governments, who consented to be interviewed even when it involved an imposition on their normal working schedules. Having promised to respect their confidences and to maintain their anonymity, I hope they will find that this book does no great violence to the ideas they shared with me.

Material support for research and travel came from a number of sources. The project was begun under the sponsorship of the Harvard University Center for International Affairs and was furthered by grants from the Social Sciences Research Council and the American Philosophical Society. I spent four valuable months in Stockholm as a guest of the Swedish Institute for International Affairs. The final stages of research were completed under the auspices of the Cornell University Program on Peace Studies, which is sponsored jointly by the Center for International Studies and the Program on Science, Technology, and Society. A conference on problems of nuclear proliferation convened by the Program of Peace Studies in Ithaca in the spring of 1971 tested a number of the ideas presented here.

I am grateful for permissions granted to incorporate into this book expanded and updated segments of some of my early articles on the proliferation question. Briefly, the articles have appeared in *Bulletin of the Atomic Scientists* (October 1969, January 1970, October 1970), *International Organization* (Spring 1970), *Co-operation and Conflict* (Spring 1970), *Asian Survey* (September 1970), Steven Spiegel and Kenneth Waltz, eds., *Conflict in World Politics* (Cambridge: Winthrop, 1971), *Cornell International Law Journal* (Winter 1972), Richard Rosecrance, ed., *The Future of the International Strategic System* (San Francisco: Chandler, 1972), and in William Kintner and Robert Pfaltzgraff, Jr., eds., *SALT: Implications for Arms Control in the 1970's* (Pittsburgh: University of Pittsburgh Press, 1973).

Any attempt to list all the individuals on whom I have imposed, for comment, earlier versions of this book is certain to be incomplete. Let me thus simply convey special thanks to a few who may have had to hear more than their share of my views on the NPT: Peter Auer, Donald Brennan, Milton Esman, Lawrence Finkelstein, Morton Halperin, Robert Jervis, Franklin Long, Richard Rosecrance, Lawrence Scheinman, Thomas Schelling, and Kenneth Waltz.

Most significant of all, the book has been enjoyable because of the encouragement and patience of my wife, Aline, who by now knows much more than she wants to know about nuclear proliferation.

THE POLITICS OF
NUCLEAR
PROLIFERATION

1

Introduction

A world in which many more countries acquire nuclear weapons could be a much less pleasant one. Wars that today kill thousands might instead kill millions; wars that now are averted might instead be launched simply because each side preemptively hastened to use its weapons, feeling that it dared not hesitate.

To prevent a further spread of nuclear weapons, the United States and the Soviet Union have offered for signature a Nuclear Non-Proliferation Treaty (NPT) freezing the number of states that can possess nuclear weapons at the current five—the United States, the Soviet Union, Great Britain, France, and Communist China. Under this treaty, countries that possess nuclear weapons would pledge not to give them away and not to assist other nations in producing them. Countries not possessing such explosives would pledge neither to accept nor to manufacture them; moreover, they would agree to accept inspection safeguards by the International Atomic Energy Agency (IAEA) on all peaceful nuclear activities in order to ensure that such facilities and materials are not diverted to the production of nuclear explosives.[1]

The United States and the Soviet Union came to agreement on an NPT text between the autumn of 1967 and the spring of 1968, and the treaty was offered for signature on 1 July 1968. Ratification by 43 states, including the United States, the Soviet

[1] The text of the Nuclear Non-Proliferation Treaty appears in United States Arms Control and Disarmament Agency, *Documents on Disarmament, 1968* (Washington, D.C.: U.S. Government Printing Office, 1969), pp. 461-65.

Union, and Great Britain, was required for the treaty to go into effect. Acceptance was delayed somewhat in the aftermath of the Russian intervention in Czechoslovakia, but on 5 March 1970 the Soviet Union and the United States joined those nations that had already ratified the treaty, and the treaty was activated.

Although the United States and the Soviet Union support the treaty, it does not follow that all other states significant to the question agree. Britain has signed and ratified the treaty, but the two other nuclear-weapons powers, France and Communist China, have not. Among the non-weapons nations that have significant nuclear industries, only Canada and Sweden had ratified the NPT by the end of 1972; West Germany, Italy, India, Japan, Australia, Israel, Brazil, and Argentina—all of which are potential producers of nuclear weapons—had not.[2]

The objections explicitly raised by these nations may be categorized as military, political, or economic.[3] They may also be classified as pretended or sincere. This book is an attempt to separate pretense from reality in national attitudes toward the Nuclear Non-Proliferation Treaty. It is hoped that it will clarify some of the analysis of prospects for arms control and also note some comparable aspects of the foreign policy processes in the several countries involved.

MILITARY ARGUMENTS

In a world in which wars are still possible, any treaty which requires renunciation of a weapon might be expected to draw

[2] Informed discussions of the spread of nuclear technology appear in the following: C. F. Barnaby, ed., *Preventing the Spread of Nuclear Weapons* (London: Souvenir Press, 1969); Leonard Beaton, *Must the Bomb Spread?* (Hammondsworth, Middlesex: Penguin, 1966); Leonard Beaton and John Maddox, *The Spread of Nuclear Weapons* (London: Chatto and Windus, 1962); Alastair Buchan, ed., *A World of Nuclear Powers?* (Englewood Cliffs: Prentice-Hall, 1966); Arnold Kramish, *The Peaceful Atom and Foreign Policy* (New York: Harper and Row, 1963); Mason Willrich and Bernard Boskey, eds., *Nuclear Proliferation: Prospects for Control* (New York: Dunellen, 1970); and Sir Solly Zuckerman, Alva Myrdal, and Lester Pearson, "The Control of Proliferation: Three Views," *Adelphi Papers*, vol. 29 (October 1966).

[3] For a view highly critical of the way the superpowers handled the treaty, see Elizabeth Young, "The Control of Proliferation: The 1968 Treaty in Hindsight and Forecast," *Adelphi Papers*, vol. 56 (April 1969).

objections. Perhaps only a few of the recalcitrant nations will openly admit to a craving for nuclear weapons, but the adoption of other arguments and excuses could still rouse our suspicion that the weapons option was the crux of the matter.

Opponents of the treaty may insist that current defense guarantees are not sufficiently reliable for them to surrender the nuclear-weapons option. Great-power "nuclear umbrellas" have been questioned ever since Pierre Gallois in 1961 professed the fear that America would never retaliate against Moscow just because she had invaded Europe, and suggested that an independent French retaliatory force was therefore required.[4]

Several questions have been raised by recalcitrant nations about the effects of surrendering the nuclear-weapons option: First, would a nuclear power protect a non-nuclear state against nuclear attack by credibly threatening to retaliate against the attacker? Would the United States or the Soviet Union drop a bomb on Peking after a Chinese attack on Tokyo, although it would probably lead to further Chinese attacks on Moscow or San Francisco? Second, does the nuclear umbrella extend to deterring conventional attacks? Would nuclear weapons be made available to stop a Chinese horde from crossing the Himalayas, or a victorious Arab army from advancing on Tel Aviv? Third, can the nuclear umbrella be used to deter acquisition of nuclear weapons by unfriendly neighbors? Can it assure Egypt that Israel will be deterred from manufacturing nuclear weapons? The five states allowed to retain nuclear weapons probably have everything to gain from acceptance of the treaty, and they openly or secretly welcome the prohibitions it imposes on other countries, which will never be able to join the "nuclear club." Conversely, the rest of the states probably worry about their loss of deterrence.

Of course there are exceptions. Some small countries will worry so little about the deterrence of invasions that they will be glad to renounce nuclear weapons. Some nuclear-weapons states will indeed be concerned whenever their smaller allies feel insecure and unprotected. States denied nuclear weapons, although they resent prohibitions imposed on themselves, may welcome prohibitions imposed on other non-weapons states. Partisans of the NPT would thus argue that the returns of a successful treaty should outweigh

[4] Pierre Gallois, *The Balance of Terror* (Boston: Houghton Mifflin, 1961).

forgone military options for reasonable nations simply because nuclear nonproliferation makes wars less likely and less horrible for all concerned. A state that desires to be sixth should realize that in forgoing this status it may prevent the emergence of a seventh.

It is common knowledge that nuclear weapons threaten mankind; yet we sometimes need to remind ourselves of the variety of disasters that such weapons make possible. Nuclear weapons can simply be used to kill larger numbers of people than any other kind of weapon; indeed there have been individuals in history who wished such homicide. If Hitler had come into the possession of nuclear weapons, would he not directly have chosen to destroy Tel Aviv? Could we have trusted Syngman Rhee or Kim Il Sung or Patrice Lumumba or Rafael Trujillo with such deadly armaments?

Limited stockpiles of nuclear weapons may be used as a catalytic device to induce much greater destruction. Instead of destroying Tel Aviv directly with his limited stockpile, Hitler (or Mao) might have obtained a much larger impact by using it to bait other nuclear forces into action. The first ("killer") approach doesn't involve motivation; the second is based on a theory of rational behavior. A state that preferred a world in which the United States and the Soviet Union no longer held a material preponderance might take a small nuclear strike at both on the assumption that it would lead each to complete the other's destruction.

Motivation aside, nuclear weapons can come into use simply to head off the use of other nuclear weapons. If our bombers or missiles carry off their "counterforce" preemptive strike with the proper degree of surprise, we can hope to keep the other side from damaging us and put aside all complicated theories about deterring them.

What will be the general strategic impact of nuclear weapons if they spread into the arsenals of country after country? Will they encourage an offensive position or will they reinforce a defensive one, with the result that any military initiative during a crisis is discouraged? Will use of nuclear weapons favor large forces or small? The answers to these questions must to some extent depend on whether we are talking about air, ground, or sea warfare.

If we take first the case of air warfare, the introduction of nuclear weapons would very probably favor an offensive position and a smaller air force. Nuclear weapons presumably enable a single bomber to destroy all the airplanes gathered on any particular airdrome. The weapons tend to serve as an equalizer by helping the air force that gets its planes into the air first to get over into the opposing air space. Thus, if a preemptive offensive is favored, nuclear weapons may make war more likely.

In the case of ground warfare, it has been conjectured that nuclear weapons will help the defense because offensive break-throughs normally require a large concentration of troops and armored vehicles, which might present a very inviting target for tactical nuclear weapons—again, nuclear weapons seemingly assist the side with fewer troops as an equalizer. Yet typically the defense must have a pretty thick wall of troops in order to force the offense to concentrate. The offense could use nuclear weapons against this thick defensive wall, since the defense almost by definition gives away its location more clearly than the offense. If attractive targets for tactical nuclear weapons appeared in both armies, the ensuing high casualties might favor the army with the more extensive reinforcements. Thus, nuclear weapons in ground warfare may often help the offensive and larger armies. If the attacker has to come through a narrow mountain pass, however, nuclear weapons might more clearly favor the defense, conceivably as preemplaced nuclear land mines. Perhaps the ideal defensive weapon would be a bomb too cumbersome to be mobile.

Finally, in the case of sea warfare, nuclear weapons might seemingly serve as an equalizer, since one bomb fired from a ship can destroy many ships. Yet the net impact might well be to aid the defense, since the offensive exploitation of sea power has often involved a well-organized and numerically superior assembly. Lanchester Square Laws and conventional wisdom have ad-monished us never to "divide the fleet"; but the threat of nuclear attack will put an end to most of the great naval formation maneuvers, as well as to massive amphibious operations.

Yet supporters of proliferation would note that nuclear weapons can be used to prevent wars as well as to wipe out long-hated enemies or to execute clever offensives. Indeed the most "traditional" use of nuclear weapons has been in the deterrence of obnoxious behavior by other states. If you use nuclear weapons against my cities, I will use them against yours.

Moreover, if you try to occupy my cities even *without* nuclear weapons, I will drop atomic bombs on your cities. Proliferation advocates would contend that the threat of massive retaliation generally deters aggression against any possessor of nuclear weapons; however, because of credibility problems, as noted above, this deterrence does not extend to allies of nuclear powers. A spread of nuclear weapons might result in the spread of a zone of peace. Lest we too readily accept this reasoning, we must note that the probability of conventional war had already begun to decline prior to this spread of nuclear weapons. While slightly increasing the impediment to regional wars, a spread of nuclear weapons might mainly change the source of the impediment; yet the cost of such wars would be greatly increased.

If we focus on nations as we know them, it is not easy to predict whether proliferation will make nuclear war more likely. The world now has five nuclear powers with many thousands of warheads, but only two warheads have ever been used (by one power) in combat. As proliferation continues, the record may not remain so good, but the ambivalence of so many nations on the halting of proliferation suggests that they do not expect the record to be drastically worsened either.

POLITICAL ARGUMENTS

At first glance, we might interpret most foreign resistance to the NPT as a reflection of a malicious craving for nuclear weapons. More honorable and generally acceptable arguments can be advanced for opposing the treaty, however; even if a nation has no intention of ever producing nuclear weapons, it might relish the legal option of producing them for the political concessions this option can produce or simply for the prestige of being on an equal legal footing with the superpowers. The treaty may thus proclaim what otherwise could have been left as a tacit agreement. If Italy had never acquired nuclear weapons, it still might have preferred not to sign a treaty which formally called for giving them up, accepting that Paris could always have such bombs but Rome never could. It is readily understandable that political prestige might make a nation wish to avoid the obligations of the NPT or even to make a bomb. Various governments might also have reserved acceptance of the NPT until some concrete political concession—the return of Okinawa to Japan or Soviet concessions to Bonn—was offered in exchange for renunciation of the bomb.

We have seen a long debate about whether the existing world of five nuclear powers has been bipolar or multipolar. One semantic resolution of this argument has been to call the *source* of political power indeed bipolar, since only U.S. and Soviet industrial and military establishments really amount to anything substantial. When we consider which *preferences* are satisfied and which are frustrated, however, we would conclude rather that the world is multipolar. In terms of freeing nations to do as they wish, the second nuclear force will balance out and neutralize the first. Thus, *two* nuclear powers may have the impact of five or twenty on the distribution of influence; everyone benefits, but the first and second are the causative powers.

Closer examination of the policies of France, Britain, and China might lead us to conclude that their status as nuclear powers has helped them gain (or retain) prestige that would not have been obtainable otherwise. In the case of Communist China, nuclear weapons may even have added some real strategic independence. Yet it is not easy to decide whether these states are substantially more able to ignore and frustrate the great powers than other non-nuclear states (for example, Japan) or whether they would have lost much of their freedom of action if they had remained non-nuclear. Further nuclear proliferation may not have much impact on the present multipolar distribution of real political influence, but it may alter the extent to which the two superpowers are causative. The possibility remains that a legal halting of proliferation would indeed entail a shifting of political influence away from the various states toward the two super-powers. Thus, as we have said before, proliferation may stabilize the existing distribution of influence; bans to proliferation may upset this distribution.

A halt to proliferation will probably now indeed require the acceptance of the NPT by those who have not yet signed. But Italy has argued that true European unification will be impossible once the NPT has endorsed France and Britain as nuclear states and relegated Germany and Italy permanently to the position of non-weapons status. Prestige is a real part of the substance of political influence; if the sovereignty of nations must be compromised tomorrow to avoid proliferation next week, the distribution of political influence may be altered tomorrow rather than next week—in the *opposite* direction.

ECONOMIC ARGUMENTS

Renunciation of the weapons options can also have economic costs. Scientists have argued that valuable spin-offs in the peaceful application of atomic energy may be lost if weapons research is explicitly forbidden, since the technologies often overlap. More importantly, the verification process that the IAEA would install to assure outside nations that materials were not being diverted to the manufacture of weapons would have definite costs. That is, the equipment and manpower required for such inspections would have monetary costs of their own.[5]

Unhappily, the line between what is allowed and what is forbidden under the NPT is not as clear as many might have hoped. Military and civilian nuclear technology are so linked as to leave almost no gap at all. Economic needs for electrical power may necessitate the building of power reactors. Some power reactors are fueled with natural uranium, which is freely available in a number of countries, and are thus not dependent on external sources of supply, which could be cut off. Others are fueled with enriched uranium, which typically must be purchased from the United States but which might easily enough be further enriched to weapons grade U-235. Either type of reactor can produce plutonium as well as electrical power. Plutonium, when the economics of power production justify it, can be further enriched in a chemical separation plant for use as fuel for various kinds of reactors. Thus, over time larger quantities of enriched uranium and plutonium, both of which can be used for bomb production, will come into circulation in various countries, and national plants for the production of one or the other will become commercially advisable.

---

[5] A basic analysis of the costs and problems of inspection appears in Arnold Kramish, "The Watched and the Unwatched," *Adelphi Papers*, vol. 36 (June 1967).

One can argue that this menacing overlap between civilian and military technology is nothing new, since responsible officials have been alert to it for years. No one expects the IAEA inspectorate to spend most of its time watching for autonomous, explicitly military programs; instead the focus of attention will be on the civilian activities which in their natural growth move ever nearer to military activities. Anyone who reads the IAEA safeguards documents, which have been evolving since 1961,[6] will be impressed by their ingenuity; there exist no major gaps through which materials might slip unnoticed into bomb manufacture. The capacity for manufacturing bombs may be growing; however, an international inspection or safeguards system can still serve to alert the outside world about a deliberately launched bomb program before it reaches fruition.

This IAEA vigilance may not relieve our concern; it may only prove how well-taken it was. In effect the ingenuity of the IAEA safeguard documents may illustrate the running of a hopeless race. The inspectorate must be alert to what is legitimate to be certain that nothing illegitimate (that is, lacking in civilian purpose or moving toward the realm of explosives) is undertaken. As more and more activities become legitimate—not only reactors but reprocessing plants, separation plants, and perhaps soon uranium diffusion plants—the inspectors would have to follow along, always watching a new layer of legitimacy, to make certain that the next "illegitimate" layer is not undertaken.

If civilian nuclear activities are legal—even if bomb production is not—they will bring a nation to within a relatively short time-lag of the bomb, perhaps only a matter of months. Why indeed should the outside world care *how far* any particular state is from acquiring nuclear weapons? We would probably prefer that it not have such weapons and we would therefore hope that it is quite far from acquiring them. If this distance were great, we could expect to detect any deliberate move to acquire bombs before it was successful and still have time to initiate political, economic, or

[6] The International Atomic Energy Agency (IAEA), based in Vienna, will have primary responsibility for inspecting peaceful nuclear installations in order to assure compliance with the Nuclear Non-Proliferation Treaty. Comprehensive accounts of the evolution of IAEA practices can be found in Kramish, *The Peaceful Atom and Foreign Policy,* and in Lawrence Scheinman, "Nuclear Safeguards, the Peaceful Atom, and the IAEA," *International Conciliation,* vol. 572 (March 1969).

even military retaliatory moves. In fact, anticipation of such detection and retaliation may deter the move to acquire the bomb. The longer the time-lag before a bomb could be produced, the more time for the outside world to notice and react to such a move, and the greater the deterrent. In the tensest situation, the sanction for any perceived grasping for bombs might be war and nuclear attack. However, sanctions far short of this might suffice, such as general political condemnation by the outside world, the freezing of assets and remittances, or an embargo of some sort.

It is very difficult to be specific about the costs of an inspection system under the aegis of the IAEA. Costs of course increase as greater security is achieved, and security declines when a more economical, less obtrusive system is adopted. As the overlap between militarily useful projects and peaceful projects increases, moreover, it will become more and more difficult to assure the outside world that no bombs could as yet have been produced by one country or another.

In addition to the problem of time wasted in accommodating inspectors, much attention has been focused on the risk of industrial espionage; IAEA inspectors might, for example, sell technical details of West German discoveries to Soviet or American industrial establishments. Discussion of the likely extent of such criminality has hardly approached a definite conclusion; technical ingenuity on automated inspection techniques may reduce the inspectorate's personal presence and visitation.

From a political point of view, it is also still possible that IAEA inspection would be very effective even if it were not technically airtight. As suggested above, the international inspector's explicit function may soon be made obsolete by the shrinking of the time-lag to bombs. Yet it would be an overstatement to say that international inspectors would serve no purpose once the crucial time-lag had fallen below six months; by their very presence, they might lengthen the physical time-lag.

To begin with a relatively trivial point, no one can be certain that a bomb program would not be detected by the inspectors as they wander around the premises. Since estimates of likely secrecy will have to be probabilistic, the addition of a safeguards system may yet deter a bomb program for fear that the outside world would be roused before the program could be completed. Even if they could not detect a bomb program any earlier than the cheating state wanted them to, the presence of inspectors would be a continual, visible reminder of the state's treaty obligations, a reminder that some form of international retaliation would follow the acquisition of bombs. A decision to violate the NPT will be a delicate decision for any state; while forgetting about international feelings might make it easier for a state to decide to go ahead, the presence of inspectors would make such forgetfulness harder.

The presence of inspectors generally would upset the legitimacy of purely national decisions on nuclear weaponry; in effect it would undermine a nation's sovereignty in the nuclear field. Crucial physicists might become confused enough in their loyalties to betray a national bomb program to a visiting IAEA representative. Moreover, the presence of inspectors would make it more difficult for outside states to forget or withdraw their threats of reprisal; the insult to outside states would be compounded if their nationals on station were ignored and defied.

Finally, the IAEA safeguards procedures, however inadequate, would force the government in question to institute national auditing and control systems which otherwise might not have been established. Because of international inspection, any move toward bomb production would require an explicit national decision, rather than be allowed to go forward by default. Thus, bomb acquisition might be forestalled in states which otherwise might have let small cliques of scientists of military men launch military weapons programs.[7]

Although we have a commonsense analysis of the possible impact of IAEA inspection, there are still questions as to how obtrusive or effective the system need be; for example, a token inspectorate which would check bomb aspirations in Japan might be very ineffective in Israel. Could the Vienna Agency plan for different degrees of certainty in two different countries, especially when physical certainty always has an economic price tag for the inspected party?

---

[7] Some of the following argument was made earlier in a broader arms-control context by Lawrence S. Finkelstein, "New Trends in International Affairs," *World Politics* 18 (October 1965): 117-26.

"NATIONAL INTERESTS" REEXAMINED

The substantive issues of the NPT are typically discussed in the terms outlined above. Yet there is reason to question whether all such arguments can be taken at face value. Even when in diplomacy the issues are correctly stated, the individual stating them often has more in mind than the simple pursuit of truth. Substantive arguments may mask more complicated conflicts of interest between states that already possess nuclear weapons and states that would be forbidden to acquire them and also important tensions and conflicts of interest within particular nations.

In cases where the issues concerning two governments are subtle, it has indeed become fashionable to analyze foreign policy in terms of interests and viewpoints below the national level. Upon closer analysis, it will often seem that the real conflict is actually a conflict between two factions within one of the governments rather than between the two governments; private business interests or personal interests of government officials must thus be sought out and identified in order to understand the positions adopted at all stages of the diplomatic process. Hidden coalitions between bureaucratic units in the two governments confronting an opposing coalition of different agents from the same two governments may emerge. Thus, there may be only a few bureaucrats who are motivated by the particular "national interest" of their countries, since various other interests absorb a goodly fraction of the attention of the men who have the decisionmaking power.

The chapters that follow focus on the particular countries, or more precisely, as the titles suggest, on the particular capitals in which the political game concerning the NPT has been played. The bias suggested by these titles is deliberate. The viewpoint that the obvious national needs and interests of India or Japan or West Germany directly determine the reaction to the NPT is rejected in this analysis. Rather, it is assumed that New Delhi or Tokyo or Bonn are also focal points of a process which sometimes has a life of its own or at least uses the physical, economic, and military ramifications of nuclear energy in much more sophisticated ways than may have been supposed to date.

The sequence of chapters is not of overwhelming significance. After Washington and Moscow, the two sources of the treaty, the sequence more or less suggests the author's feelings about which countries will present obstacles to the treaty. The conventional wisdom has often seen Bonn as a primary source of trouble; this book's opposite conclusion is thus underscored. Paris, Pretoria, and Peking have little in common except that each can be a giver of nuclear weapons (South Africa only of the uranium) and none has endorsed the treaty.

Many of the impressions conveyed in this study were gained from interviews with individuals in and out of government in the countries discussed. A great number of other impressions, some of which are accepted and others of which are not, are drawn from previously published works on the NPT.

2

Washington

A proliferation problem could have been defined from the day the first atomic device was exploded, since Americans might naturally have wanted to keep the weapon for themselves.[1] Following the use of two bombs against Japan, President Truman in fact announced that the United States would not share its bomb "secrets." The United States Atomic Energy Act of 1946 transformed Truman's statement into law, thereby excluding even those nations whose physicists had assisted in the wartime American bomb project.

The 1946 Baruch Plan could be considered the original NPT; whether or not it would have terminated American options on nuclear weapons, it could have served to foreclose them for everyone else.[2] Skeptics can contend that the plan would always have supported the American monopoly of bomb-production expertise and was therefore defective and unacceptable to the Soviet Union; but this contention is really based on the argument that one and only one state could have the bomb if there were no "proliferation" and that it was thus too early for an NPT. Perhaps two nuclear powers are preferable to one, as well as to three or twenty, in terms of preserving peace and stability.

---

[1] A more comprehensive discussion of American attitudes about the NPT can be found in William B. Bader, *The United States and the Spread of Nuclear Weapons* (New York: Pegasus, 1968).

[2] The text of the Acheson-Lilienthal proposals as presented by Baruch can be found in United States Arms Control and Disarmament Agency, *Documents on Disarmament, 1945-1959* (Washington, D.C.: U.S. Government Printing Office, 1969), vol. I, pp. 7-76.

From 1945 to 1958, the world was seemingly divided into two camps; after a time, very few nations that mattered did not belong to one of the two camps. The detonation of an A-bomb by Russia in 1949 thus paradoxically terminated much of any anxiety about proliferation, since the worst damage had seemingly been done. A British capacity to produce nuclear devices before 1949 might have drawn American opposition because of the increased risk of leaks of technological information to the Russians or because the British precedent could be used by Moscow to demand such information. But with the genie out of the bottle and the bomb "in enemy hands" by the time Great Britain tested her first nuclear device in 1952, friendship and memories of British assistance on the World War II bomb project reduced American opposition to Great Britain's membership in the nuclear club.

Fears of leakage of advanced technological information to the Russians persisted, of course, because the H-bomb had yet to be developed and tested by either the United States or the Soviet Union. More extensive proposals for the sharing of weapons with Great Britain in 1950 and 1951 were balked by congressional opposition because of fear of Soviet espionage; and assistance to France was similarly precluded until the late 1950s because of suspected Communist sympathies on the part of prominent French physicists.[3] Yet the nuclear question, as a "proliferation problem," receded once "each bloc" had the bomb. Prospects of a German atomic bomb might have roused great fears on the part of both superpowers—greater in fact than their fears of each other—but in 1953 the Germanies were not yet even permitted to rearm conventionally. Nuclear weapons in the hands of an irresponsible regime might also have instilled fears, but who in his right mind was going to give nuclear weapons to Syngman Rhee or Kim II Sung?

It will always be difficult to decide whether the decisions of France and China to produce nuclear weapons preceded or followed their moves toward political independence of Washington and Moscow. As a result of their bomb production and moves toward independence coming so close together, the nuclear problem was again regarded in terms of proliferation. The late

[3] See H. L. Neiberg, *Nuclear Secrecy and Foreign Policy* (Washington, D.C.: Public Affairs Press, 1964), for a full discussion of pre-NPT American reasoning on controls on nuclear information.

1950s saw France launch a nuclear program which by predominantly indigenous competence would produce an A-bomb in 1960;[4] by 1959 a combination of Soviet assistance and indigenous talent similarly launched China on a program which would produce an A-bomb in 1964.[5] Although it was too late to halt China and France, other possibilities of weapons spread might soon have to be confronted, at least in part because of the precedent set by these fourth and fifth nuclear-weapons states. The United States in the 1950s had begun to base its defense of the NATO area on a program which included German rearmament and tactical nuclear weapons; over time this program would probably raise the question of German access to nuclear weapons. The United States had also launched an "Atoms for Peace" program, which encouraged the development of nuclear energy, making available to many states their first nuclear reactors and fissionable materials.[6]

It is easy to understand why the United States (and the Soviet Union) might become quite concerned about proliferation in the mid 1960s, much more so than in the days of bipolar political arrangements and limited world expertise in nuclear matters. The Nuclear Non-Proliferation Treaty illustrates a very straightforward approach to the problem of nuclear-weapons spread. Being a coauthor of the NPT, the United States might be expected to wholeheartedly support it; yet the treaty's history (and indeed its current status) shows some serious ambiguities in the American position. What is the American position on the NPT? Is America in favor of the treaty or not? Is every American not opposed to proliferation?

THE COSTS OF THE NPT

Almost every American is to some degree opposed to the further spread of nuclear weapons; the question is, How much?

---

[4] See Lawrence Scheinman, *Atomic Energy Policy in France under the Fourth Republic* (Princeton: Princeton University Press, 1965).

[5] Arnold Kramish, "The Great Chinese Bomb Puzzle—and a Solution," *Fortune*, vol. lxxiii (June 1966).

[6] The background and consequences of the "Atoms for Peace" program are discussed in Arnold Kramish, *The Peaceful Atom and Foreign Policy* (New York: Harper and Row, 1963).

This question may be serious; for it is likely that the great majority of the American public have never consciously contemplated the problem and indeed do not even know that "NPT" refers to a Nuclear Non-Proliferation Treaty. The NPT has drawn an uneven reception among more active participants in the political process—negative side-effects accompany any attempt to halt nuclear weapons spread by means of a formal treaty.

The possibility has always existed that a treaty such as the NPT would prejudice American options for deployment of nuclear weapons. Nonproliferation and nondeployment seem to be linked psychologically, even if the NPT distinguishes quite clearly between the two. To take a simple example, before the NPT any local resentment of a U.S. deployment might have been assuaged by a token exposure of local forces to nuclear armaments; a treaty may make this impossible. Since the mid 1950s, the deterrent to a Russian conquest of Western Europe has depended heavily on the threat that aggression would quickly lead to all-out nuclear war. The alternative of a purely conventional defense would have been more costly and would have required longer terms of military service for both Americans and Europeans. Stationing American nuclear weapons in Germany has been an effective way of making the threat of all-out nuclear war credible, which explains much of the Eisenhower administration's aversion to nonproliferation and denuclearization treaty proposals.

Related to the above was the possibility that the treaty might generally help Communist states more than it would help U.S. allies, if only because the Soviet Union endorsed the nonproliferation principle before the United States, with the result that U.S. acceptance seemed to be a vindication of Soviet moral judgment. Just as discussion of the treaty seemed to endorse Soviet slanders of U.S. allies, so might signing the treaty. Communist propaganda and Communist gains aside, U.S. allies might have been unhappy about a nonproliferation treaty for a host of other reasons, good or bad. If so, the United States would not be indifferent to their feelings about foreign policy issues, just as it would expect them not to be indifferent to its feelings. Apart from allies, other friendly countries have complained voluminously about the treaty; indeed there has been little enthusiasm for the NPT outside the United States and the Soviet Union, and the typical American is likely to be skeptical of a treaty pushed forward against the wishes of so many countries whose governments Americans respect.

Among the more serious objections advanced against the treaty has been the argument that its inspection provisions will impose economic costs on the countries required to submit to safeguards. The inspectors may be obtrusive and may waste the time of valuable indigenous technicians; or the injunction against bomb production may inhibit the full development of peaceful nuclear technology in the country. Americans generally have looked to technology to find solutions to most if not all of the world's problems; thus, artificial inhibition to technology would inevitably cause some disquiet. Disquiet would also be caused by any suspicion that other states may demand compensation from the United States and other nuclear-weapons states for opportunities lost as a result of renunciation of nuclear weapons. However helpful it might have been in selling the treaty, the United States was not going to commit itself to more generous policies of foreign aid or even to sharing commercially valuable secrets about nuclear technology.

In the late 1950s, nonproliferation proposals would have been viewed with suspicion in Washington because they threatened to erode American commitments to West Germany and the NATO area in general. In the late 1960s the United States would have been concerned as to whether the treaty presupposed new commitments to nations renouncing nuclear weapons in order to restore their defensive security to what it might have been if they had become the sixth or seventh nuclear power. After Vietnam, few Americans will be inclined toward defensive commitments in Asia or even the Middle East.[7]

At times it may be difficult to assign priorities between arms control and other foreign policy goals; yet there is a need to choose among methods of arms control, no matter what its priority. Spelling out an abstract issue of "proliferation" may have masked some other important trade-offs for American policy. The world does not consist of "nth" countries, but rather of Germany, India, and so on. If the likelihood of one country going nuclear is reduced, the likelihood of another may as a result be increased; this raises a new need for choice.

---

[7] For the relevant portions of Secretary Rusk's testimony denying any new U.S. commitment as the result of the NPT, see U.S. Senate, Ninetieth Congress, Second Session, Foreign Relations Committee, *Hearings on the Non-Proliferation Treaty* (Washington, D.C.: U.S. Government Printing Office, 1968), pp. 15-16.

It is not yet certain how the treaty will function here; it may serve to chain together all the prospective nuclear-weapons states, since abstract discussion of a problem makes the abstraction real. Thus, any sixth nuclear-weapons state may indeed open the floodgates for the seventh through the fourteenth, if only because the great powers will have been successfully challenged and proven to be paper tigers. The first state to acquire nuclear weapons now will have done so in face of the well-articulated and well-deployed opposition of the United States; this was not true for Great Britain or France or even for the Soviet Union or China. Thus, Sweden must realize that its atomic bombs would not be seen simply as the result of a Swedish military decision but as a precedent, which should deter Sweden or any other erstwhile sixth. The appearance of a sixth is perhaps less likely because of American support for the NPT and for the antiproliferation principle and, in part, because a seventh would probably follow closely behind a sixth.

It is also generally probable that any state which contemplates and then rejects the treaty will have thereby whetted its appetite for weapons. Although the treaty may effectively inhibit the states which sign and ratify it, it may accelerate decisions in favor of weapons by other states. If there were no treaty, pressure against nuclear weapons could be tailor-made for each nation, with the result that each decision would not be regarded as a worldwide precedent. Thus, the decision of one more nation to join the nuclear club would have had less earthshaking implications, and the atomic-bomb question might never have come up in countries where political or technical circumstances otherwise would not have caused it to be raised.

This raised a different kind of issue for American policy on a nonproliferation treaty, an issue of technique rather than of priority. A subtle approach might offer many advantages over explication. If the United States opposed weapons spread subtly, without so openly admitting to it, it might more effectively inhibit this spread and, at the same time, reassure allies, avoid the costs of inspection, and so on.

In a sense, President Eisenhower favored the more flexible approach. The Republican administration was most probably opposed to an unlimited spread of nuclear weapons, but it seemingly assumed that an unadvertised opposition to proliferation would be the most effective approach. Whether the United

States would assist France and under what terms such assistance would be given was generally left unclear. At times the Congressional Joint Committee on Atomic Energy forced the administration to be more explicit than it had intended because Congress was still wary of nuclear sharing because of the chance of a leak to the Russians.[8] The Eisenhower administration avoided a doctrinaire position when it could, however, and no endorsement of anything like a nonproliferation treaty was ever forthcoming.

For an illustration of what some foreign observers interpret as more subtle American antiproliferation policy, one can turn to the Atomic Energy Commission's prices for enriched uranium. The AEC's prices were low enough to tempt electrical power producers in a number of countries to become dependent on the enriched-uranium fuel cycle rather than build power reactors fueled with natural uranium. Whether or not the prices were artificially low, one can see this as an effective subversion of the use of natural-uranium-fueled reactors, which could have done double duty as nuclear-weapons programs. The legal tie of a dependence on American fuel can be of much longer duration than the tie of the NPT. Legally, withdrawal from the terms of the NPT simply requires three months' notice; the bilateral agreements between the United States and recipient nations require that the recipient nation virtually start from scratch if it wishes to manufacture weapons, since its entire current fuel cycle is dependent on enriched uranium from the United States.

THE GAINS OF THE NPT

The American monopoly on fuel and technology could not continue for very long, and some countries (for example, India and Israel) have been careful not to be seduced onto the enriched-uranium path. Although one can tick off a series of plausible American reservations about the NPT, there might have been enough powerful arguments in favor of the treaty to overcome all of them; the problem of weapons spread might have become urgent.[9]

[8] See Neiberg, *Nuclear Secrecy and Foreign Policy*, pp. 180-95.
[9] Examples of the American consensus thinking on the need for an NPT can be found in *Stopping the Spread of Nuclear Weapons* (New York: United Nations, 1967), and

Some sentiment for a more explicit barrier to further national nuclear forces was expressed in the earliest days of the Kennedy administration, especially by persons who had studied arms-control problems in an abstract and analytical manner. As early as 1958, the National Planning Association had published a book entitled *1970 Without Arms Control*,[10] in which it was predicted that from ten to fifteen countries would have nuclear weapons by 1970 if no artificial barriers to proliferation were imposed. Arguments influential within the Department of Defense under Secretary of Defense McNamara were presented quite explicitly in Albert Wohlstetter's 1961 article, "Nuclear Sharing: NATO and the N + 1 Country."[11] Given the general desire of President Kennedy and Secretary McNamara to "put the nuclear genie back into the bottle," a move toward an explicit nonproliferation treaty would seem quite appropriate; indeed, one argument in favor of the Test-Ban Treaty of 1963 was that it promised to make proliferation more difficult.

The Kennedy administration's outlook on proliferation was colored by its views on military strategy as a whole. The Republicans had been inclined to rely on tactical nuclear weapons in forward deployments as a deterrent to a war in Europe, thinking that escalation would be so likely if war came that the Russians presumably never would launch one. The Democrats now were inclined to remove nuclear weapons from the front lines, or at least to create an option of limited conventional war (for example, by installing permissive action link [PAL] devices to control the use of tactical weapons). This would mean increasing the conventional ground forces for the defense of the NATO area, but the risks of small engagements escalating into a nuclear World War III would be reduced.

Closer controls on American nuclear weapons were fully consistent with more explicit barriers to new national nuclear forces. There were thus several nuclear genies that the Kennedy administration wished to put back into the bottle, and these at least superficially seemed to be related to each other. European

---

in the *Final Report* of the International Assembly on Nuclear Weapons (Scarborough Conference), reprinted in *Survival* VIII (September 1966), pp. 278-81.
[10] National Planning Association, *1970 Without Arms Control* (Washington, D.C., 1958).
[11] Albert Wohlstetter, "Nuclear Sharing: NATO and the N + 1 Country," *Foreign Affairs* (April 1961), pp. 355-87.

resistance to the higher taxes and longer draft term required by a conventional defense came soon enough, and the Democratic administration was in the end forced to deploy even more tactical nuclear weapons to West Germany.[12] Yet, the United States never quite gave up the hope of keeping nuclear weapons out of a battle, which inevitably brought the proliferation question into the open on both sides.

The only major countertrend to such explicit resistance to proliferation was the Department of State's lobby for a Multilateral Nuclear Force (MLF).[13] The scheme was essentially a carry-over from some vaguer nuclear-weapons-sharing proposals that had emerged at the end of the Eisenhower years, which had first involved land-based missiles, then Polaris missilefiring submarines, and then missiles based on vessels resembling merchant ships. The firing of Polaris missiles from such a force of mixed-manned surface ships would have been subject to the veto of the United States and other NATO members; hence it did not constitute proliferation, since it really only imposed a new safety catch on some of the American arsenal. By exposing Germans and other naval personnel to such weapons, however, the MLF hopefully would satisfy whatever nuclear aspirations Germany might have. Critics of the MLF pointed to the obvious lack of substance in the nuclear-weapons-sharing involved and suggested that over time this would lead Germany to demand a more real proliferation. What had been intended to satisfy imagined tastes for nuclear weapons might rather stimulate such tastes.

The NPT and the MLF could thus be seen as overlapping, noncontradictory approaches to the prevention of a further spread of nuclear weapons. Yet many observers doubted whether they could indeed be kept psychologically reconciled. Most important was the reaction of the Soviet Union, which consistently denounced the MLF proposal and insisted that any nonproliferation treaty specifically ban such multilateral forces. Since the United States had become committed to the MLF and had induced the Bonn regime to become similarly committed, no progress toward an agreed NPT text seemed likely. It was the introduction of the

---

[12] Institute for Strategic Studies, *The Military Balance: 1965-66* (London, 1966), p. 13.
[13] For a good account of the various sides to the MLF debate, see Arthur M. Schlesinger, *A Thousand Days* (Boston: Houghton Mifflin, 1965), pp. 850-56.

Social Democrats into the German government through the Grand Coalition of 1966 that allowed the MLF finally to be gracefully abandoned; the superpowers could then move ahead to a nonproliferation treaty.

By 1967, the Soviet Union and the United States had come close to agreement on an NPT text; compromise on only an inspection clause remained. Explicit approval or disapproval of a multilateral force would now be avoided because the United States had ceased to endorse MLF proposals. If the Western European states should ever unite politically and then move toward a multilateral nuclear arrangement, the superpowers might have to debate the meaning of the treaty text they had drafted; if such a situation does not arise, this debate can perhaps be infinitely postponed. When agreement on the inspection clause was reached in June 1968, the treaty was offered for signature. The NPT was important not only because of its substance; it was the first actively cooperative venture between the United States and the Soviet Union since 1945. There have been other instances in which one of the superpowers desisted from sabotaging or undermining a venture of the other, but the history of the NPT has seen the two states cooperating and even conspiring together. This new cooperation at Geneva had drawbacks of its own; yet, happily, it proved that such cooperation was possible, and it established a bureaucratic group with a vested interest in continuing it. The conduit of cooperation herewith opened was to be useful in more serious matters—in talks on the limitation of strategic weapons, for example.

One can see why the treaty had such strong support in the United States from President Johnson and particularly within the Arms Control and Disarmament Agency (ACDA) of the Department of State. Given the relatively abstract nature of much of U.S. arms-control thinking both in and out of the government, the idea of a nonproliferation treaty would certainly be attractive. On the surface, the intent of the treaty was clear and unexceptionable: it was arms control, it might reduce the costs and likelihood of war, and given a willingness to evade the MLF issue, it was attainable. For President Johnson, the treaty in 1968 might have served to somewhat counterweight the Vietnam war by showing that the administration could achieve some form of cooperation with the

Communist world. After President Johnson left office, the ACDA's support for the treaty remained strong, as strong as that in any other part of the American government.

BUREAUCRATIC CONSIDERATIONS

Although the ACDA is solidly in favor of the treaty, the support of other sections of the Department of State has been more lukewarm. Some of these are long-time partisans of European unification who see the NPT as a threat to Euratom and thus to the entire "functional" international process. Others are desk-officers who for years have had to deal with specific, rather than "nth," countries, who are hence specially sensitive to the political feelings of the particular states reluctant to sign the treaty.

Similar ambivalences are to be found in other relevant parts of the American government. The Atomic Energy Commission is certainly not in favor of spreading weapons and has been a conscientious supporter of safeguards systems; yet it has also been a booster for peaceful uses of atomic energy, including peaceful nuclear explosions. AEC personnel in the United States and in the field are not all of one mind concerning the effect international inspection might have on the full development of nuclear electrical power. For example, the AEC has a tremendous backlog of experience in inspection techniques; working from this backlog, it might well have concluded that Euratom's inspection systems are technically equal to or better than those of the IAEA and that the objections of Brussels to a full takeover by Vienna thus deserve support. Periodic explosions that are part of the AEC's "gas-buggy" program suggest that peaceful explosions can be used to tap mineral deposits deep underground; what is more important, they strengthen the resistance to the NPT of countries which wish to retain this option for themselves.[14]

The Joint Committee on Atomic Energy of the U.S. Congress (JCAE), which has had extraordinary influence during its short

[14] See David B. Brooks and Henry R. Myers, "Plowshare Evaluation," in *Nuclear Proliferation: Prospects for Control,* ed. M. Willrich and B. Boskey (New York: Dunellen, 1970), pp. 87-101.

career, may be an influential source of support for the NPT. The committee favors the treaty and the halting of weapons spread, but it most probably will balk at any use of American fuel supply to pressure Germany and Italy to sign the NPT. Statements by individual members of the committee (for example, Representative Hosmer)[15] at times suggest a more traditional approach to antiproliferation policy, namely, that weapons simply be forgone as part of a general alliance and commercial cooperation with the United States. Whether Germany formally renounces nuclear weapons (for the second time) is much less important, in this view, than keeping Germany closely involved technologically and politically with the United States.

PRESIDENT NIXON

The most important influence on NPT decisions (albeit not always the decisive influence) comes from the president himself. President Johnson has been replaced since 1969 by President Richard Nixon, a man much less openly enthusiastic about the treaty. The differences in the two men's attitudes can be variously explained. As a candidate for office in 1968, Nixon would naturally have been prone to criticize projects of the Democratic incumbent, even if he had ambivalent feelings about their substance. After the Russian invasion of Czechoslovakia, it was appropriate for a Republican candidate to take a tough line against Communist aggression and to imply that the Senate should not rush to ratify a treaty that had been coauthored with the Soviets. The Eisenhower approach to proliferation matters had been generally milder and less openly doctrinaire; Nixon could thus see the NPT as a serious mistake in approach, as compared with the approach a Republican administration would have taken.

Nixon's choices in 1969 could obviously not be as free as they might have been in 1965. The United States had invested considerable national prestige in the NPT. A rejection of it might thoroughly confuse most nations and anger as well as confuse the Soviet Union. Although proliferation might have been halted by

[15] See Hosmer statement in U.S. Senate, Ninetieth Congress, *Hearings on the Non-Proliferation Treaty*, pp. 166-67.

quiet diplomacy (itself a disputable proposition), the debate about the NPT in many countries had already inferentially opened up a debate on nuclear weapons as well; one could not turn the clock back to 1965 in India or Japan or West Germany. Even from a standpoint originally critical of the treaty, it would be better to keep the NPT alive in order to at least recover part of what might have been a poor investment. President Nixon, after some initial silence, thus announced in March 1969 that he would indeed ask the Senate to ratify the NPT.

Yet, in the opinion of the United States, there have been, and will continue to be, some issues of priority or stress concerning the treaty, issues which will not be resolved in quite the same way as they might have been before 1968. In order for the NPT to receive the required Russian ratification, a West German signature had first to be obtained. Moreover, much would depend on how that signature was extracted, that is, whether it was obtained in a manner embarrassing and derogatory to Bonn or in a manner that protected West German prestige. Deposit of the American ratification was thus in fact postponed for twelve months, until a time when German and international politics had facilitated a West German signature which would not be read as a plea of "guilty" to Russian propaganda charges. The United States under Nixon has been willing to delay ratification of the treaty in order to bolster Bonn's dignity and prestige.

Now that the treaty has gone into effect, the Nixon administration must interpret the specific legal obligations of the United States. Perhaps it will decide that the treaty calls for a termination of fuel supply when individual countries can not reach agreement with the IAEA on inspection procedures; perhaps not. If such issues should arise, the Republican response would probably be less sympathetic to the Vienna Agency.

Similarly the United States must decide how seriously and actively to pursue its Plowshare program for peaceful uses of nuclear explosions. Allegedly on the grounds of budgetary constraints, the Nixon administration has slashed funds for this program for two consecutive years. Apart from cost considerations, such slashes should, in the end, make it easier to sell the NPT. A few countries (for example, Australia) may feel that failure of the United States to pursue the AEC's research is nothing more than a denial of nuclear explosion services to those who need them for projects such as harbors, canals, and underground fuel storage cavities. In general, however, a slowing of American research will mean a slowing of publicity about all the alleged benefits such explosives have in the civilian sector. Opponents of Plowshare argue that all such benefits are, in any event, greatly exaggerated, since problems of residual radioactivity are very difficult to solve; foreign proponents of Plowshare regard such opposition as deriving mostly from fears of proliferation, not from objective analyses of the material possibilities. Whether or not civilian benefits can, in time, be extracted from peaceful nuclear explosions, the world generally still clings to the intuitive view that all nuclear explosions must be dangerous. From the standpoint of winning acceptance for the NPT, it may be best to let the world continue to cling to such a view.

More important issues of stress may remain. How much pressure should and will be applied to secure ratifications from countries which have signed the NPT? The president, in conversations with Japanese, German, and Australian leaders, does not seem to accord this issue much priority. What retaliation, if any, will be imposed on nations which not only refuse to sign the treaty but move to "violate" it by manufacturing nuclear explosives? In the cases of Israel and India, less pressure is being applied than might have been expected in the past.

We can rationalize that an absence of American pressure is optimal for the success of the treaty, since most forms of pressure would be counterproductive. The Nixon administration thus can have its cake and eat it—it can do "all it can" for the NPT by assuming a soft-sell low posture and save its more explicit trading points for use in twisting arms on entirely different issues, such as the Japanese textiles issue. There is indeed something to be said for the quiet approach. The treaty has been signed by a great number of states, most of which normally would not fail to ratify a treaty within a reasonable length of time after signing. Lebanon, Jordan, and Ecuador have, after suitable delays, ratified the treaty. Their ratification did not necessarily come as the work of Washington or Moscow; international law, with its tradition and logic, supplies certain pressures of its own. Yet this trend has hardly proven that the quiet approach is the best approach the

United States could take in order to push the treaty along. At times, if it were willing to sacrifice other goals, the Nixon administration probably could do more.

SALT AND THE NPT

Since the United States and the Soviet Union presented their text of the NPT, the next significant development in arms control has assuredly been the Strategic Arms Limitation Treaty (SALT) negotiations between these two superpowers in Helsinki and Vienna. Naturally, some interrelationship between these two consecutive arms-control steps of major importance has been expected, since the final success or failure of either venture is not yet assured and because the success of one could contribute to the success or failure of the other.

The immediate impetus for the SALT talks stemmed from increased Soviet procurement of offensive and defensive anti-ballistic missile (ABM) systems. Heading off a new round of the arms race between the superpowers offered substantial advantages for the two states, including financial savings, an improved political atmosphere, and reduced risks of a World War III. Such considerations had indeed made the Johnson administration receptive to the idea of such talks prior to the Russian invasion of Czechoslovakia, which then, however, made the gesture seem inappropriate. The Nixon administration avoided any display of enthusiasm for the SALT idea for a time, but by the autumn of 1969 it had agreed to go ahead with such Soviet-American negotiations.

The negotiations obviously will have some impact on the prevention of further nuclear proliferation and on the acceptance of the NPT. Yet other considerations, such as those cited above, have more directly prompted Washington and Moscow to begin the talks. The links with proliferation may thus be weak and secondary in importance; perhaps even worse, the direction of any linkage is confusing.

All along there has even been an argument that arms limitations such as those achieved in the past round of the SALT talks would actually weaken the chances of the NPT. In this view, a continued arms race between the United States and the Soviet Union is necessary in order that incentives for other states to enter the nuclear club be reduced and the neighbors of such states be reassured. Thus, the elimination of antiballistic missiles and the latest missiles from the American and Russian arsenals might make a Japanese nuclear program look more appropriate, since it would now be easy for Tokyo to "catch up." A continued arms race between the two superpowers has indeed cast doubt on the military efficacy of the British, French, and Chinese nuclear-weapons programs, and even more so, on the hypothetical programs of India and Israel. According to this view, if India wanted assurance that it could not be blackmailed by China, it would have been to India's advantage for Russia and the United States to go ahead with ABM protection for their cities, so that the threat of nuclear reprisal by one or both on behalf of India would still remain credible.

By the same line of argument, any great-power decision to go beyond a freeze, that is, to reduce nuclear-weapons arsenals, would simply further tempt an erstwhile sixth or seventh nuclear power to enter the club when the standards of admission and annual dues had just been lowered. Total nuclear disarmament by the five existing nuclear-weapons states might seem to be the full moral equivalent of what the NPT proposes for all the others; yet the chances of adherence to an NPT by all the others would then probably be considerably less than they are today.

Yet there is, of course, a strong counterargument that successful SALT negotiations would help assure the success of the NPT, since Article VI of the NPT itself imposes an obligation on the great powers to "pursue negotiations in good faith on effective measures relating to cessation of the nuclear arms race at an early date and to nuclear disarmament, and on a treaty on general and complete disarmament under strict and effective international control." Many commentators on the NPT have questioned the apparent asymmetry and inequity of the treaty, since it tolerates forever the nuclear-weapons prerogative of those five nations that were prudent enough to enter the club before 1967—the United States, the Soviet Union, Great Britain, France, and China—but denies that prerogative to all other nations. Any failure to reach agreement may thus be regarded by some countries as an excuse to withhold or, even worse, to withdraw ratifications of the treaty.

Disarmament has perhaps too long been viewed as a moral question, portrayed as such by opposing sides in the cold war as each tried to direct world accusations against the other. Yet, as long as much of the world clings to this moralistic interpretation of disarmament, a failure of the SALT talks may well be bitterly regarded by those nations which, according to the NPT, must renounce nuclear weapons; their reaction may upset full or final agreement on the NPT.

If the United States and the Soviet Union are serious now about preventing further nuclear proliferation, it might also seem plausible that success would depend on how well they can coordinate their efforts. SALT in itself may be enormously important; it might be worth pursuing a Soviet-American agreement on SALT even if such an agreement had no impact on proliferation. The most important impact of a SALT agreement may well be to serve as a reminder to the two superpowers (and to the rest of the world) that the issues on which they can agree may be more important than those on which they compete.

A successful SALT negotiation might elevate the officials on each side who staked their reputations on great-power cooperation and thereby preserve and enlarge conduits for agreement on other subjects. A successful experience in Helsinki and Vienna would be immensely valuable training for American and Russian diplomats posted in South Asia and the Middle East. Concerted and coordinated action by the United States and the Soviet Union in these regions could conceivably slow down nuclear proliferation and, at the same time, bolster the "conventional warfare" peace. Ideally, each of the powers would carefully apply as much leverage as it could muster to deter a decision in favor of bomb production; and each would also do what it could to reassure India, Israel, Japan, or the Germanies that renunciation of nuclear weapons will not inevitably lead to military or political disasters.

This may only illustrate the difficulty of maintaining or exploiting any expected spin-off from SALT; there are other very real issues which other sectors of the American and Soviet bureaucracies are committed to contesting for a while longer. It is always difficult to predict whether a first step in détente will generate anything else. Will a successful SALT agreement spill over into fuller Soviet-American coordination in the Middle East and India, or will it simply free funds to expand the navies and conventional air forces which fly superpower flags in those areas? How large will the fleets be that the two powers deploy in the Mediterranean? And how large will the deliveries of tanks and jet aircraft be to Arab states or to Israel? How clear will various pledges of support in South Asia be, and how large will the arms deliveries be there? Support by one of the great powers might make India or Israel feel secure enough to forgo, or at least postpone, any decision to go nuclear. Yet, if the sixth country is determined to become the sixth nuclear power, too full support by a great power may remove whatever checks and leverage the superpowers have in deterring proliferation. Fuller coordination between the United States and the Soviet Union will thus not be so easy to achieve. SALT may fail; if it succeeds, its spin-offs may be outweighed by other events.

Even if the United States and the Soviet Union move ahead with a closer cooperation based on détente in Western Europe and real progress on SALT, the format of such cooperation may become so menacing to Israel and India that they will decide to become "nth" countries. Thus, it may be an oversimplification to say that closer cooperation between Washington and Moscow will be the most important impact of the SALT talks on the prevention of proliferation.

For example, the United States could obviously work more closely with the Soviet Union in the Middle East, by giving greater support to Arab grievances and applying greater pressure on Israel to withdraw from the territories seized in 1967. Perhaps this coordination would come about as a by-product of SALT and an American desire to deemphasize military commitments. Yet these moves certainly would not postpone an Israeli decision to produce a bomb. A closeness produced by Russian moves in the opposite direction might have a more beneficial effect, at least in terms of stopping nuclear proliferation.

When the United States opened contacts with the Peking regime in 1971, it declared that in doing so it had no hostile intent toward the Soviet Union or anyone else. If the Soviet Union responded in the same spirit, it might be able to improve its own relations with Communist China; perhaps Soviet proposals for a broadening of the SALT talks in order to include Peking—as well as London and Paris; that is, to bring in all the nuclear-weapons states—were a straightforward illustration of such a response. A broadening of the talks would illustrate fuller coordination and less counterproduction between the two superpowers—or among all five of the nuclear powers—but it would hardly reassure India or make less likely any Indian decisions to produce atomic bombs.

India, Israel, and other states may each have particular substantive defense problems which remain to be explored. It would be a happy world if all forms of Soviet-American détente, or Chinese-Soviet-American détente, also served to ease the tensions of the Middle East and South Asia and to decrease the attractions of nuclear weaponry. Perhaps greater compatability of goals can be found eventually; in the meantime, some contradictory goals will persist.

3

Moscow

The Soviet Union has favored a ban on the spread of nuclear weapons for more than a decade—longer than the United States. An explicit nonproliferation provision was included in the Soviet Memorandum on Partial Measures, which was transmitted to the U.N. General Assembly in September 1957,[1] and the Soviet Union supported the Irish General Assembly resolution opposing proliferation in 1958. For an even longer time, the Soviet Union has been committed to measures seemingly related in spirit to halting proliferation: a complete ban on the existence or use of nuclear weapons anywhere (since about 1945); a halt of all testing of nuclear explosives (since 1955; the Russians indeed launched a unilateral moratorium on such test detonations in March 1958); and the designation of various nuclearfree zones, geographically defined regions in which nuclear weapons would neither be used nor deployed (since 1956, including particularly the Polish Rapacki Plan of 1957).[2] All of these schemes would have made the spread of nuclear weapons to additional countries more difficult or impossible. Soviet statements at times acknowledged this constraint and welcomed it.

[1] The documentary record of earlier Soviet proposals can be found in United States Arms Control and Disarmament Agency, *Documents on Disarmament, 1945-1959* (Washington, D.C.: U.S. Government Printing Office, 1969) (hereinafter cited as *Documents on Disarmament, 19—*). The text of the Rapacki Plan appears on pp. 889-92.
[2] For an interesting analysis of Soviet intentions in formulating various arms-control proposals, see Lincoln Bloomfield, Walter Clemens, and Franklyn Griffiths, *Khrushchev and the Arms Race* (Cambridge: The M.I.T. Press, 1966).

SUSPECT MOTIVES

If accepted, each of the schemes would indeed have had many additional effects, which would have been much less acceptable to the United States and its allies and which have led the West to doubt the sincerity of the Soviet commitment to halting proliferation per se. To the end of the Eisenhower years, it was generally assumed that the Soviet Union wanted to delegitimize any and all use of nuclear weapons, primarily because of U.S. nuclear-weapons superiority and because the defense of Western Europe seemingly depended on the threat of escalation to this form of weaponry. Soviet proposals to limit membership in the nuclear club were thus normally entangled with bans on any use of nuclear weapons, even by members of the club. When not explicitly conjoined, such Soviet proposals on nuclear weaponry were almost as a matter of course misunderstood to be overlapping and logically related; and they were implicitly presented as such in Soviet disarmament propaganda.

The Eisenhower administration may indeed have been reluctant to see the number of states holding nuclear weapons extend beyond the three states which held them in 1952, the United States, the Soviet Union, and Great Britain. Yet Soviet proposals for an explicit NPT necessarily had to be viewed with suspicion as long as the Republicans could not consider any bans on the use of nuclear weapons in defense of Western Europe. It would be some time before a halt to proliferation would not suggest a no-first-use obligation on the part of the states already holding nuclear weapons. Once the Russians, by weapons acquisitions and strategic declarations, had accepted escalation to nuclear warfare for certain circumstances, an NPT might be viewed in a different cold-war context.

There was also a problem concerning inspection or verification, which was the core of so much Soviet-American disagreement on other forms of disarmament. The United States had consistently contended that outside inspection should be required to induce compliance with the terms of any disarmament agreement. The Soviet Union had almost as consistently responded that this would really amount to espionage and that an agreement negotiated in good faith should be executed without policing by foreign inspectors. Despite the differences between halting of proliferation and other disarmament, the verification dispute would likely enter into the proliferation problem, since to deny or affirm a need for inspection here would seemingly vindicate positions taken elsewhere.

This contest on inspection was illustrated in a series of Irish antiproliferation resolutions presented to the U.N. General Assembly between 1958 and 1960. The United States showed its opposition by abstaining on the resolutions in 1958 and in 1960, in each case citing as its reason the omission of verification in the resolution text. The 1959 resolution, which referred to transfers of "control" rather than of "possession," seemed to be tailored according to American two-key policies; it thus drew the support of the United States and the abstention (that is, the opposition) of the Soviet bloc.

Consistent with these positions, the Soviet Union repeatedly opposed IAEA inspection operations until 1963. As in the case of other forms of disarmament, the Soviet Union might have been sincerely in favor of halting nuclear proliferation, unless it required extensive international inspection in order to certify that each side was carrying out its part of the bargain. The NPT might have served as a self-policing measure for the superpowers if neither wished to give nuclear weapons to its allies, but could the Soviet Union indeed vouch for all of its satellites? One can argue that the Russians were seeking excuses with which to convince the Chinese that there was no choice but to deny them nuclear weapons. Yet what besides international verification would have prevented Peking from moving ahead, just as it did, to acquire its own nuclear arsenal?

As support for the argument that international inspection was indeed necessary, there is evidence that the Soviet Union was not especially careful to impose strict safeguards on the reactors and fuels it delivered to other nations in the late 1950s.[3] Some of these were research reactors which handled relatively insignificant amounts of uranium and plutonium. (A precedent of inspection parallel to the procedure the United States was insisting upon, could have been set here but it was not.) Some of the reactors,

[3] See Arnold Kramish, *The Peaceful Atom and Foreign Policy* (New York: Harper and Row, 1963), pp. 64-65, 179-82.

such as those exported to Communist China, were of more significant size.

It is also apparently true that materials accounting even within the Soviet Union was not as strict as that of the AEC inside the United States, which makes it feasible that individual plant managers could easily have diverted fissionable materials without the central authority being aware of it. Some of this apparent laxness can of course be attributed to the existence of party hierarchies and controls within the Soviet Union which supplement those nominally part of the state—a plant manager is not known to the authorities in Moscow only by the books he keeps. This kind of supplemental authority would not really be effective outside the Communist bloc, however, or even outside the Soviet Union within the bloc, after the demise of Stalin. The failure to insist on safeguards in China, Egypt, or Czechoslovakia must thus also be traced to the general propaganda line Moscow had taken on any and all inspection related to disarmament, namely, that it was unnecessary among honorable and peaceloving nations.

PROLIFERATION FROM MOSCOW

It is difficult to believe that China's progress toward production of A-bombs and H-bombs was not accelerated by aid received from the Soviet Union prior to 1960. It is also unlikely that the Soviet Union intended Peking to have such bombs. The first Chinese bomb utilized a uranium core, rather than the expected plutonium, and therefore foreboded a rapid progress toward production of an H-bomb. According to one theory, the Soviet Union supplied China with not only power reactors, which use uranium as a fuel and produce plutonium, but also an enrichment plant, which prepares uranium for use in reactors (or in bombs).[4]

The Chinese success must be attributed in some part to the Soviet Union's careless gift of aid which could readily be used for military purposes, which was given in the hope that the Communist alliance (by which the Soviets might perhaps maintain enough leverage to veto any Chinese bomb decisions) might yet

[4] See Arnold Kramish, "The Great Chinese Bomb Puzzle—and a Solution," *Fortune* (June 1966).

thereby be preserved. A more flatly negative answer to Chinese requests for assistance would have been quite costly for Moscow. A flat "no" to India would similarly have had a price; and Russian criticisms of IAEA inspection at this stage seemed to be responsive to New Delhi's desires for freedom of action rather than directed toward achieving a sure barrier to proliferation.

The Soviet Union must have been aware of the fact that it could also be costly for the United States to give its allies a negative answer. It is entirely plausible that Moscow was seeking to drive the United States to accept this cost while itself escaping any similar costs vis-à-vis China or India. The propaganda gains for the Soviet Union of a campaign against nuclear proliferation were relatively clear. At the least it might cause tensions within the NATO alliance and cast doubt on the legitimacy and credibility of the tactical use of nuclear weapons in defense of Europe. As long as the inspection issue was ducked, relations with Communist and nonaligned powers would not be damaged very much. Had the United States chosen to agree to no inspection rather than accept the propaganda losses, Bonn would have been slightly discredited and Moscow would have had a card to play as it chose against Chinese requests for nuclear assistance. Willingly or not, the Soviet Union clearly accelerated China's move toward bomb production.

Proliferation of a very different sort was involved in the deployment of Soviet missiles to Cuba in 1962. We might suppose that the Russians would not have relinquished control over such weapons to Castro, just as we have not relinquished control to Turkey or West Germany. Yet the reaction of the United States was at least in part due to the implicit fear that firing control might pass into the hands of the Cubans. It is always difficult to be sure that weapons deployed among a satellite's troops will not come into the satellite's control. One can design elaborate devices—indeed the United States has done so—that would make firings difficult if, for instance, Cuban or West German soldiers overpowered the representatives of the great power that owned the weapons.[5] We do not know if such devices were present on the missiles deployed to Cuba.

The Kennedy administration's extreme reaction to the deployment of missiles to Cuba illustrated an aversion to Cuban nuclear

---

[5] See Joel Larus, *Nuclear Weapons Safety and the Common Defense* (Columbus: Ohio State University Press, 1967), pp. 80-86, for a full discussion of the PAL device.

armament which might have existed all along. Kennedy had been much more explicitly averse than Eisenhower to any general spread of nuclear weapons and was more willing to pay some price to put the nuclear genie back into the bottle. In part, this willingness was derived from new doubts on whether the prospect of nuclear escalation in a European war really benefited the United States; it also came from the fact that the existence of Soviet missiles and H-bomb tests had moved the Soviet Union into a position where it might no longer be suspected of seeking to delegitimate all use of nuclear weapons. In the aftermath of Cuba, and even before, Moscow could thus have been sure that the Democratic administration would give more support to a formal ban on proliferation. President Kennedy's attitudes had been revealed in the 1961 revised U.S. General and Complete Disarmament Plan, which included a nonproliferation clause, and in the U.S. vote for the Irish Resolution in the U.N. General Assembly that year.

THE TEST BAN

The Nuclear Test-Ban Treaty, which followed the Cuban missile crisis, came well after the termination of any Russian nuclear assistance to Peking but before the first Chinese nuclear detonations. The development of a joint Soviet-American position on the test ban has been slow and complicated and only partially related to halting the spread of nuclear weapons.[6] As with antiproliferation treaties, the test-ban proposals at first seemed intended to compromise the U.S. prerogative to use nuclear weapons and, thus, the American deterrent. It was clear that agreement to a total test ban at any point would ipso facto have mobilized world opinion against "nth" nuclear powers; yet the informal moratorium entered into by the United States, Great Britain, and the Soviet Union from 1958 to 1961 did not dissuade France from detonating its first bomb in 1960. The final Test-Ban Treaty of 1963 was seen as a possible hindrance to China's entrance to the

---

[6] A full account of the issues and processes involved in negotiating the test ban can be found in Harold Karan Jacobson and Eric Stein, *Diplomats, Scientists, and Politicians* (Ann Arbor: University of Michigan Press, 1966).

club, although an underground detonation would have been perfectly legal if Peking had chosen to sign. As it was, Peking denounced the treaty as an example of Soviet-American collusion.[7]

The Test-Ban Treaty was most importantly a signal of Soviet-American joint interest in avoiding additional proliferation and, after it was signed by most of the world, an imperfect barrier to such proliferation. Had the formal treaty prohibited all explosions, rather than simply explosions above the ground, it might have approximated an implicit nonproliferation treaty. It would have been theoretically possible for nations to design and assemble bombs without any test detonations, but tests are still perhaps the only conclusive way to convince the world that one indeed has the bomb. As it is, a country like India can enter the club today without violating its treaty obligations, simply by keeping its test detonations below the earth's surface.

The occasion of the Test-Ban Treaty apparently saw some signals of further interest in halting proliferation on the part of the United States. However, American suggestions of possible joint military efforts to preempt Chinese production of nuclear weapons were rebuffed by the Soviet Union.[8] It was apparently either too early or too late to ask Moscow to contemplate war against Peking in conjunction with the Americans, even for the purpose of stopping nuclear proliferation.

MIXED MOTIVES

After the Soviet Union's open break with Peking, the Soviet position on nuclear proliferation seemed to become much clearer between 1963 and 1967: there would be no question of offering nuclear weapons to any Communist state and pressure should be applied in order to make the United States promise not to give them to its allies. No longer having anything to lose by the treaty, the Soviet Union could hope to keep West Germany, as well as such nonaligned states as India or Israel, from obtaining the bomb.

[7] See, for example, "A Comment on the Soviet Government's Statement of August 21," *Peking Review* VI, September 6, 1963, p. 9.
[8] *New York Times,* October 2, 1964, p. 3.

If Cuba or Rumania wanted the bomb, there would not be any indecision (as there had been with China) as to whether it could be denied them. Indeed, Russian signature on a nonproliferation accord could be conveniently cited as an excuse for diverting any such requests.

Western commentary on Soviet commitment to the NPT has centered around several abstractions. The first, which could be labeled cold-warrior skepticism, is the argument that Moscow obviously supports the treaty wholeheartedly because, in the classical cold-war context, it has everything to gain and nothing to lose. The pill is not a bitter one for Moscow, because the Soviet Union did not intend to give nuclear weapons to any such allies as Hungary or Czechoslovakia; rather, the NPT is a clear victory in that it denies the United States the right to supply such weapons to Bonn or the NATO Multilateral Force (MLF). If it thereby embarrasses Bonn, so much the better.

Yet this argument somewhat oversimplifies the Soviet Union's calculus of its costs and gains from the treaty. Although the pill is not bitter for Moscow, it is bitter enough for Cairo, New Delhi, Havana, and many other capitals; and forcing a bitter pill on someone else can itself be a bitter experience. Once Moscow chose to endorse the NPT publicly, it thus had to share the resentment which otherwise might only have been directed against Washington and London. The Soviet Union could have emulated France, hoping that proliferation would not occur but refusing to accept the onus of endorsing concrete steps to block it. In cold-war terms it has sacrificed some of its political position by openly endorsing the treaty.

Given that Soviet support for the treaty transcends a simple cold-war calculus, a second set of commentators have adopted the abstraction that Moscow is indeed as fully and abstractly committed to halting proliferation as Washington. Events outside Europe could easily explain a serious Russian commitment to the halting of proliferation everywhere. China detonated its first nuclear device in 1964 and, as a result, clearly captured the attention of persons influential in Indian nuclear development. In 1965, India and Pakistan went to war over Kashmir again, and the Soviet Union found itself offering its good offices to negotiate a peace at Tashkent. Continuing reports emerged of nuclear research in Israel based on French assistance. In 1967 Israel defeated the Arab states in another round of war in the Middle East. Whatever the Soviet intentions had been with regard to the European strategic balance, nuclear-weapons production was increasingly possible now in both the Middle East and South Asia; and the Soviet Union most probably preferred that this possibility not materialize in areas such as these, which were so prone to armed conflict.

Yet such an analysis can exaggerate Soviet commitment to the treaty. The idea of delegitimizing nuclear weapons had become outmoded, since the Soviet Union indeed approached parity in nuclear strength, but other cold-war considerations would still remain important. Having for so long denounced arms-control inspection as espionage, the Soviet Union still found it embarrassing to express enthusiasm for it. More importantly, it seemed likely that the West German regime might not commit itself to an NPT. The treaty could contribute to the East German Communist regime's strength, which was dependent on disparagement of the Bonn regime, by phrasing the NPT proposal in a way that would stress Bonn's evil revanchist qualities. Stressing the application the treaty might have for Germany might further reduce the chance of damage to the Soviet position in India, since Moscow could take less than full responsibility for administering the bitter pill. At all times, even up to the very present, the Soviet Union has thus still had to decide whether to sincerely advance an agreement that both superpowers are likely to want or to exploit the American desire for an NPT in order to make gains in the cold-war context.

Doubts about the sincerity of the Soviet desire for an NPT persisted even after the Test-Ban Treaty of 1963, because of a seemingly continued Russian opposition or indifference to inspection as part of any ban on proliferation. The first American NPT draft in 1965 included a requirement for "application of IAEA *or* equivalent international safeguards";[9] conversely, the Russian draft which followed made no mention at all of inspection safeguards.[10]

[9] *Documents on Disarmament, 1965,* pp. 347-49.
[10] Ibid., pp. 443-46.

PROPAGANDA AGAINST BONN

The two drafts differed equally about future establishment of any multinational nuclear forces, which clearly suggested that the Soviet Union was adapting the NPT issue as a means of baiting Bonn. The only multinational force under consideration in the mid 1960s was the NATO Multilateral Force (MLF) which would exist according to a scheme, primarily originated by the U.S. Department of State, which would give Germans and others a sense of participation in nuclear matters without also giving them the capability for firing nuclear weapons without American approval. Few countries in Europe had shown any real enthusiasm for the MLF, which had come to mean surface ships with mixed crews carrying Polaris missiles. West Germany had committed itself mostly in response to American urging. The U.S. NPT draft explicitly allowed such a multilateral force; the Soviet draft explicitly forbade it. Russian attacks on the MLF centered on charges that it would give Bonn its own nuclear capability.

It might indeed be unfair to attribute Soviet opposition to the MLF simply to a desire to put anti-German propaganda ahead of progress on halting proliferation. As proposed, the MLF offered Germany only a safety catch on some of the U.S. arsenal and no finger on the trigger; rather than creating any new options, it constricted the options for use of some nuclear weapons. A large number of American nuclear weapons were already in place (and today continue to be) in West Germany under two-key arrangements not really different from those called for by the MLF. Yet the U.S. statements on the MLF were not resolute in asserting that the United States would always retain its veto. While no presidential statement ever said as much, statements of State Department officials at times hinted that a joint European control body might someday assume firing control over the MLF and that the United States might give up its veto. This body would surely include Great Britain and France; thus, there would still be no "proliferation." Yet the pace of the general political momentum (as opposed to the logical) would have suggested further veto relaxations until Bonn might someday have asked for the right to fire without approval from any nuclear power. Indeed, the mere proximity of German sailors to the warheads of the MLF suggested the emergence of such authority; the MLF might thus have whetted appetites which would then remain frustrated until real proliferation took place.

Yet, as mentioned above, German soldiers and airmen already had been placed in similar contact with American nuclear weapons, and no great demand for more total access had emerged. The Russian fear of proliferation via the MLF may have been real, but it was also illogical; it could just as easily have been exaggerated, self-serving propaganda. The Soviets could blacken Bonn's reputation, whenever the precarious Pankow situation required it, by charging that Germany lusted after nuclear weapons.

The West German elections of 1966 brought the Social Democrats (SPD) into the cabinet in the Grand Coalition; the Russian choice was thereby eased considerably because the SPD had been opposed in principle to nuclear weapons for Germany. The new government in Bonn could thus come off the MLF limb onto which it had been coaxed by the United States, and the United States could drop all emphasis on the multilateral option without causing its ally too much embarrassment. The change in German political outlook was not the result of Russian pressure, but it served to make further pressure less necessary or profitable. Since Germany no longer talked about the MLF, the Soviet Union would not have reason to fear proliferation for quite a while; nor could it continue to erode Bonn's reputation by harping on the multilateral option. The joint NPT draft presented late in 1967 thus neatly finessed the MLF question by means of vague wording, which each side might interpret differently if the multilateral question should ever seriously arise again.

Although the MLF issue had been defused, it seemed that the Soviet Union had not yet lost interest in engaging Bonn in a discussion of the text of the NPT. By the end of 1967 the Russians had completely reversed their position on inspection; now they were insisting on IAEA safeguards for all nations renouncing weapons, even in cases where the United States might have been willing to substitute equivalent forms of control. If it had not been for the special problems concerning West Germany and Euratom, the move might have been seen as a generous concession to earlier Western positions. This inspection still would not take place within the boundaries of the Soviet Union, so that cynics could claim that Moscow had not conceded any of its total exemption from inspection. Yet IAEA safeguards would now take effect within all of the Russian satellites; outsiders from the Vienna Agency would now have access to Bulgaria, East Germany, Rumania, and so on.

Yet the major problem was that Euratom had been allowed to inspect itself for almost ten years according to a system which its members found quite satisfactory. Skeptics feared that Moscow was using the treaty for the disparagement it would bring the West German regime, and they cited the enthusiastic Russian endorsement of IAEA inspection as evidence. According to this view, the United States had fallen into a trap because it had not supported the special status of Euratom strongly enough. There were good reasons to uphold Euratom, in that it was a proven, nonobtrusive control system which supported momentum for European unity; however, when Germany presented such arguments, the Russians could again accuse them of wanting bombs and wanting to avoid a nonproliferation treaty.

Some of the Russian dogmatism about IAEA authority may have reflected a desire for additional propaganda mileage at the expense of the German Federal Republic. Yet Soviet aversion to exemptions from IAEA inspection may not have been very unreasonable; indeed, it was somewhat shared by the American delegation to Geneva. If special treatment had been granted to Euratom, Japan could have demanded similar treatment. All the members of Euratom are also members of NATO. The United States would not have reacted with enthusiasm to a parallel inspection organization made up of only Warsaw Pact members (although it might by now have trusted the Russians to see that none of their satellites acquired weapons). Imagery can cause problems even when the images lack reality—on either side of the Iron Curtain; it would not always be apparent to an outsider that Germans were not about to dominate the decisionmaking process of Euratom.

Having become embarrassingly "more Catholic than the pope," that is, having adopted the American position on inspection more vigorously than the Americans themselves, the Russians could not yet come to agreement with the United States on an NPT draft. The draft of August 1967, which was the first draft to be agreed upon, conspicuously lacked an article on inspection; it was not inserted until January 1968. The wording of the compromise Article III left unsettled a number of the issues concerning Euratom and IAEA authority; in general, it required all signatory non-weapons nations to negotiate a safeguards agreement with Vienna but allowed that this negotiation could be handled "either individually or together with other states" (that is, Euratom). A final compromise between Vienna and Brussels thus remained to be achieved. Until it was, there was a chance that Moscow would come down too strongly on behalf of the Vienna Agency. Yet Russian pressure of late has not suggested any lack of interest in getting Bonn and its Euratom partners to accept the NPT as it stands.

THE NPT, FROM SIGNATURE TO RATIFICATION

Since the presentation of a joint NPT draft, Moscow has seemed to regard the treaty as more than a vehicle for propaganda against Bonn. Obvious Soviet pressure was brought to bear in order to persuade almost all the Arab states to sign the treaty in July 1968, immediately after it was formally offered (Algeria was the only exception among states receiving aid from the Soviet Union, and Saudi Arabia also refused to sign). Since Israel did not sign, Soviet pressure did not immediately induce any Arab ratifications; the mere signatures were clearly unpopular in the Arab capitals, which consumed some of the Soviet Union's leverage.

The Soviet Union has similarly made clear its desire that India and Japan sign the treaty, even though it has had no guarantee that, in doing so, it would accomplish any more than antagonizing individuals in those countries.[11] Soviet spokesmen have quite abrasively denied any arguments against the NPT brought up by

[11] A very useful cross-section and analysis of Soviet public statements on proliferation can be found in Roman Kolkowicz et al., *The Soviet Union and Arms Control: A Superpower Dilemma* (Baltimore: The Johns Hopkins Press, 1970), pp. 70-115. For an early Soviet discussion of the treaty specifically mentioning India and Japan, see the broadcast text reprinted in "Soviet Comments," *Survival* IX (May 1967), pp. 150-51. For later comments similarly explicit, see A. Alexeev, "Non-Proliferation Treaty and Security," *International Affairs* (Moscow) (January 1969), pp. 10-14, and A. Alexeev, "Non-Proliferation Treaty and the Non-Nuclear States," *International Affairs* (Moscow) (March 1969), pp. 9-13.

spokesmen of recalcitrant nations. Thus, Moscow has denounced as erroneous any charges that IAEA inspection would be too costly, too discriminatory, or too troublesome for non-weapons states to bear. While Brazil and other states touted the special properties and advantages of peaceful nuclear explosives, Moscow quite consistently declared that they were indistinguishable from military bombs.[12] If the Soviet Union had played a simple cold-war game, the United States might have been left to rebut such attempts to make the treaty ineffective and Moscow would only have had to address itself to the alleged threats of German nuclear revanchism.

While Russian interest in the Nuclear Non-Proliferation Treaty is still most often explicated in terms of revanchist Germans, Russian spokesmen now never deny the importance of everyone else signing. Some Soviet commentators seem to be delegated primarily to discussing Germany, and thus they often denounce Bonn; others, who seem to be just as clearly delegated to discussing proliferation and disarmament, have been relatively abstract and straightforward in asserting that any and all further proliferation is undesirable. The Russians must be credited with accepting a significant part of the onus of selling the treaty to countries such as India and Japan. Russian spokesmen (like American spokesmen) neither seek out nor run away from Asian examples when describing what the NPT is intended to prevent. India, Moscow says, should not acquire nuclear weapons and should accept the treaty; so should Japan and so should everyone else.

There were still some Soviet conflicts of interest concerning the NPT between 1968 and 1970. If Moscow had wanted to make a maximum contribution to the acceptance of the NPT, it could have voluntarily opened some or all of its peaceful nuclear facilities to IAEA inspection, as the United States and Great Britain have done. From a strictly logical point of view, such a gesture is not meaningful; IAEA inspectors hardly need to subject such facilities to strenuous or costly inspection, since the superpowers have no need to produce bombs clandestinely in peaceful facilities. However, it might have had some beneficial effect in countries which did not fully perceive the logical non

[12] *Documents on Disarmament, 1967,* p. 147.

sequitur and which would have welcomed such "sharing of the inspection burden" by all the superpowers.

The rejection of IAEA access by Moscow must thus be attributed to Soviet aversion to inspection, which has persisted since arguments with the United States on disarmament in the 1950s. Within the Soviet Union (if no longer within Rumania or Poland), inspection smacks of espionage—of threats to the military security and strength of the one great socialist state which has the responsibility for defending all other socialist states.

What will remain unclear is whether this residual distrust of external inspection serves to make the Soviet Union sympathetic to near-nuclear nations, which distrust inspection for only slightly different reasons. Will the fact that the Soviet Union talked for so long about military espionage make it sympathetic to West German anxieties about commercial espionage? Or, after Bonn has ratified the treaty, will the Russians be tempted to lobby for very thorough inspection in order to embarrass Germany and obstruct its industry?

THE NPT WITHIN THE SOCIALIST CAMP

Open Russian support for the treaty antagonizes not only national governments; it also makes local Communist parties hostile. For example, The Japanese Communist Party (JCP) has come out in opposition to the NPT. Thus the Liberal Democratic government in Tokyo is attacked by Communists on two sides; it is attacked by Moscow for not endorsing the treaty more quickly and by the JCP for not rejecting it.

The Russian stand has of course drawn the fire of Communist China, which consistently has denounced the NPT in the same terms as the test ban—the product of Soviet-American collusion, unrelated to real disarmament. The Chinese stand probably influences the Japanese Socialist Party (JSP) in Tokyo even more than it does the JCP; indeed, the JSP has also denounced the treaty. Chinese attitudes naturally explain most Albanian denunciations and abstentions concerning the NPT; yet one also encounters disquiet about the treaty at the opposite end of the Communist spectrum. Yugoslavia has signed and ratified the NPT, but it has expressed serious reservations about what seem to be the unequal sacrifices of nuclear and non-nuclear states under the treaty's terms.

Representatives of the Communist states of Eastern Europe except Albania were brought along to sign the treaty on the day it was offered. Rumania surprised a number of commentators by signing the treaty on the first day, after having criticized it quite extensively only a few months earlier.[13] Simple considerations of style might have suggested waiting a few weeks as a way to show its displeasure. The Russians had not yet invaded Czechoslovakia; therefore Rumania might not have felt a very strong fear of invasion. Rumania's behavior still serves as an index of Russian seriousness about the treaty. Moscow apparently signaled that it was taking the NPT much more seriously than Bucharest had expected, such that continued needling on this question would indeed seriously worsen relations. Ceasescu was apparently warned to pick a somewhat less sensitive issue to show his independence of Soviet leadership. Rumania had good reasons to object to the treaty; by objecting, it could show independence of Moscow and respect for Peking and also play the general game of small members of each alliance standing up to their grand patron. Since the Soviet Union had limited leverage in the Rumanian case, it must therefore have attached some high priority to the NPT in order to expend this much leverage on Rumanian signature.

Mongolia was the only Asian Communist state to sign the treaty, which illustrates the need for these states to balance Peking's friendship with Moscow's but hardly proves a lack of serious intent on Moscow's part.

Cuba did not sign the treaty. Considerations about appeasing Peking play some role here also, but Cuba, unlike Rumania, does not face any threat of Russian invasion. The "antiproliferation" ball got rolling to some extent after Khrushchev's submission to Kennedy's antiproliferation demands concerning Soviet missiles in Cuba. It is understandable that Castro would not care to enshrine this bit of history. Cuba may have been close to becoming the fifth nuclear power in 1962; one does not celebrate being a near miss.

Cuba has also refused to sign the Treaty of Tlatelolco, which

[13] For statements of the Rumanian position, see *Documents on Disarmament, 1968,* pp. 33-35, 167-71, 383-87.

was purported to establish a nuclearfree zone over all of Latin America.[14] The relation of the treaty to nonproliferation is complicated. United States adherence to the treaty's Protocol II has been based on the understanding that it retains the right to transport nuclear weapons through this area. It would have been difficult for Cuba to endorse a treaty which seemed to give the United States everything it wanted but denied for all time Cuba's right to the weapons of 1962. The Soviet Union has also refused to endorse the treaty, possibly as a concession to Cuba. Yet, the Soviet Union also objects to the treaty because it tolerates the production of "peaceful nuclear explosives," which the Soviets maintain are indistinguishable from bombs. The Soviet refusal strengthened the Soviet line against nuclear proliferation, but it may have cost the Soviet Union some Latin American good will.

One might think that the United States would assume that Moscow is now more committed to halting proliferation than in the past. Yet there will always be limits to a nation's commitment to any one part of its foreign policy; thus, other nations will always be suspicious. When Soviet intervention in Czechoslovakia was a prospect during the spring and summer of 1968, it was widely recognized that such intervention would indeed be a setback for the NPT because it would give opponents of the NPT one more excuse to defer committing themselves to the treaty. What better example could there be of a great nuclear nation pushing around a small nation which the NPT forbids ever to be nuclear?

Russian intervention thus showed that Moscow's interest in advancing the treaty did not have priority over all other considerations. Moreover, the ensuing burst of anti-West-German propaganda, with its allusions to "rights of intervention" under Articles 53 and 107 of the U.N. Charter, did not strengthen the hands of supporters of the treaty in Bonn.[15] Italy, Switzerland, and several other states which might have been about to sign the treaty chose to postpone their signatures. American ratification of the NPT was postponed, and thereafter ratification by the Soviet Union and by the Communist satellites in Eastern Europe was also postponed.

[14] For the Cuban position on the Treaty of Tlatelolco, see *Documents on Disarmament, 1967*, pp. 238-39, 538-39.
[15] For references to Articles 53 and 107 of the U.N. Charter, see *New York Times*, September 20, 1968, p. 1.

This hardly proved Russian interest in the NPT to be transitory or illusory. Given the world's disapproval of the Czech affair, the Soviet Union might have been inclined to adopt a posture of low visibility for a time, until the issue had blown over. At the Conference of Non-Nuclear Weapons States in Geneva at the end of the summer of 1968, however, the Russians lobbied as briskly as before to try to head off resolutions which might damage the treaty's prospects. In short, at a time when the costs were higher than usual, the Soviet government continued to twist arms on behalf of the Soviet-American treaty. The setback of the NPT timetable, which came as a result of the Soviet intervention in Czechoslovakia, could have been fatal for the treaty. It was not, however, and Moscow retained Prague and the NPT too.

RATIFICATION AND AFTER

The Czechoslovakia affair set the stage for a continued Russian confrontation with Bonn; but the emphasis now shifted to the timing of ratifications and/or German signature. Soviet ratification of the NPT was first postponed because of the clear unwillingness of the United States to ratify it before the 1968 election and then because of the uncertainty of President-elect Nixon's commitment to the treaty. Also, since Bonn had not yet signed the treaty, the threat remained that West Germany would follow a Russian ratification with a signature "with reservations," which might be phrased in terms of special exemptions from IAEA safeguards or other terms unacceptable to Moscow. It would have been difficult for the Soviet Union to pull out of a treaty it had already ratified, if the NPT had thereby already gone into effect; deposit of Soviet ratification would thus come only after West Germany had signed the treaty.

For the superpowers, as for everyone else, the great issues of the NPT—delicate issues of interpretation which may upset or bolster the treaty—are perhaps yet to come: How much nuclear fuel or equipment can a party to the treaty sell to a nonparty without demanding acceptance of IAEA safeguards? How far can a non-weapons state go in developing and testing bomb designs without violating the treaty? What kinds of political retaliation will apply to a country which, having rejected the treaty, also "violates" it by becoming the sixth nuclear power?

If the Soviet Union is at all ambivalent about these questions, it is only because the first test case will likely be India. In principle, Soviet commentators favor a very tough great-power interpretation of the NPT with regard to moves toward production of explosives: Research on weapons design by non-weapons states should not be tolerated, even though the treaty does not strictly forbid it and some countries—for example, Sweden—have in fact already completed most of their research. Countries which do not sign the NPT should not receive assistance on peaceful nuclear projects as readily as signatories would, although again the treaty does not legislate such discrimination. Nuclear assistance certainly should not be given to a nation which is using its indigenous nuclear resources to make explosives, even if the assistance itself remains subject to IAEA safeguards.[16] (This also exceeds the requirements of the NPT.)

Soviet arms-control experts willingly express these views most resolutely in the abstract, but a discussion of concrete instances introduces some equivocation. Clearly, the above attitudes would apply if Brazil tried to manufacture nuclear explosives. Approval of an Israeli move toward bomb production would be unthinkable. Yet when the question concerns India, the discussion is likely to take into consideration India's special strategic position vis-à-vis China and Pakistan; and it thus becomes clear that Moscow has not yet decided whether the abstract issue of proliferation outweighs concrete political considerations in this case.

It would not be difficult for an outside observer to conclude that both superpowers are slowly becoming resigned to an Indian nuclear explosion, since they know that they could never fully trust each other to stick to a coordinated tough line where the friendship of such a populous country was at stake. The NPT unfortunately defines only five memberships; Indian explosives could thus undermine the entire NPT system, especially if Japan or Australia or Pakistan chose to follow. Dissuading the Indians will be a difficult task for the superpowers, one that might fully challenge their political commitment to the halting of proliferation. Almost any other "nth" power would draw more vehement

[16] *New York Times,* February 17, 1968, p. 9.

resistance from both the Soviet Union and the United States; thus it is plausible that the greater effort will now be devoted to decoupling any seventh membership from an Indian move to nuclear weapons.

If Indian nuclear weapons were to be tolerated—it is hardly certain, from the viewpoint of New Delhi or anywhere else— Moscow would be no less unhappy about proliferation in Germany, Japan, Israel, or Brazil. Nor is there evidence of any inclination to tolerate nuclear weapons in the hands of states currently receiving Soviet military assistance.

The Soviet Union in the end may not force Cuba to sign the NPT, or Egypt to ratify it, but the treaty will hardly be meaningless for Soviet relations with these countries. Since the Soviet Union is party to the treaty, any significant peaceful nuclear assistance must be subject to IAEA inspection; and further, the Soviet Union stands pledged not to hand over nuclear weapons. If Israel gets the bomb and Egypt asks Moscow to supply a matching capability, the Soviet leadership has merely to read Cairo the treaty to explain why the request can not be granted. So also would be the case with Cuba after the acquisition by Brazil of a "peaceful nuclear explosive." The Soviet Union could of course withdraw from the treaty to please its allies in these cases, but that would let West Germany off the hook; Soviet spokesmen have noted quite specifically that Havana and Cairo will have to accept the presence of IAEA inspectors if they want reactors now. The external inspection that Castro rejected after the missile crisis is—in a modified form, to be sure—now the price of nuclear electricity.

If the Soviet Union has been at all lax in the past in its control of the uses to which its nuclear assistance has been put, there are definite limits to any laxness in the future. East Germany and other Socialist countries with reactors will not be encouraged or allowed to build reprocessing plants for purifying plutonium to what is weapons grade. If commercial considerations dictate, the plutonium can be shipped to and from the Soviet Union for reprocessing; moreover, until commercial uses for breeder reactors arise, purified plutonium will be stored in the Soviet Union, rather than in the countries "owning" it. One possible exception to the general requirement for IAEA safeguards is the use of atomic energy by the military for something other than bomb production—nuclear propulsion plants for naval vessels, for example. In the West, the Netherlands and Italy have projects for such vessels; however, the Soviet Union does not intend to encourage or allow Poland or Bulgaria to invest in nuclear-powered submarines or frigates, since their tactical need can be denied.

If one looks objectively at nuclear threats to the Soviet Union today, one might conclude that Germany should be secondary to another potential "nth" country, Israel. Since the Soviet Union has allied itself with the Arab states, it has made plausible scenarios in which Soviet forces might actually become involved in combat with Israeli forces; and Israel is obviously close to a real nuclear-weapons capability. Is not Israel, then, more of a nuclear threat for Moscow?

Yet this somewhat exaggerates the real rather than the symbolic in Soviet perceptions of military threat, for nothing is likely to cause the Soviets the anxieties that Germany does. Israel is still a small country and therefore a more difficult one for the Soviet man on the street to take seriously. Despite what Arab nations say, the Soviet Union is not committed to the eradication of Israel. Israel did not invade the Soviet Union in 1941 and it does not threaten an important Soviet satellite by its example and its nationalist appeal. Soviet fears of German nuclear armament have always been political fears as much as arms-control fears, and thus the Soviet Union will continue to see proliferation in an exaggeratedly German context. If Israel persists in its refusal to sign the NPT and thus induces Arab states to withhold ratifications, the Soviet Union will strive to keep nuclear weapons out of the Middle East and will count on like pressure from the United States. However, if Bonn had rejected the NPT, much more of the treaty's life might have been at stake.

Bonn's signature of the NPT, and the larger détente of which it is a part, has thus eased some Russian concerns as well as removed some temptations for propaganda. German participation (along with that of Great Britain and the Netherlands) in the centrifuge production of enriched uranium would previously have drawn charges that Germany intended to use the uranium for bomb production. So similarly would German sales of power reactors to Argentina or Brazil. As IAEA inspection is applied to Germany and the rest of Europe, Russians who previously took such charges seriously will feel a little reassured. Continued tenure of the Social

Democratic regime in Bonn will obviously enhance this feeling; conversely, the triumph of any right-wing coalition might upset it.

Yet the Soviet Union is definitely concerned about any prospects for European nuclear integration, for example, whereby French and British nuclear forces would come under some integrated political control in which West Germany was a participant. Legally the NPT seems deliberately obscure on this possibility, and the authors of the treaty probably hope that nuclear integration is far enough away so that the issue will not arise any time soon. As long as the use of nuclear weapons can still be vetoed by French or British governments, one can indeed argue that no proliferation would occur. Yet, in this decade, anything resembling German participation in the firing of a European nuclear-weapons system is likely to arouse Soviet hostility.

The most important unsettled question on the NPT may in the end concern the inspection arrangements signatory states must conclude with the IAEA. Part of the IAEA's arguments with countries like Germany will pertain to recognition of any special status for Euratom. Most of the argument, however, will shift to how thorough inspection should be; for although intensive inspection procedures are more reliable from an arms-control point of view, they are more burdensome economically from the point of view of country. This will be an argument for Germany, for Japan—for most near-nuclear countries. The IAEA's stand will depend partly on the pressures it perceives from the United States and the Soviet Union. On the basis of past anti-Bonn propaganda, one might have feared that the Soviet Union would insist on extremely thorough inspection to assure that no German bombs were clandestinely being assembled in the Black Forest. On the basis of the antiinspection tradition of the Soviet Union, however, we might forecast a much more reasonable approach.

In principle the Soviet Union is quite unenthusiastic about enormous staffs of inspectors diluting the sovereignty of countries. The need for reliable inspection to inhibit nuclear weapons spread has been acknowledged, but it is hoped that automation and technical developments can reduce the need for direct human access to civilian nuclear establishments. The Soviets, like the West Germans, do not believe that the number of inspectors should increase as a linear function of the amount of electricity produced.

It has thus been difficult for Moscow to take a clear position on

German suggestions that a new kind of "black box" be located at "strategic points" to reduce the human component of IAEA inspection. As long as Moscow remains suspicious of Bonn and Euratom, it will view such arguments as pleas for special exemptions, as German efforts to win a concession of legal principle which would excuse them from the supervision other nations must endure. As such suspicions are reduced, however, Bonn and Moscow may not remain so clearly at odds on this issue.

4

New Delhi

India will not sign the NPT unless international politics undergo some drastic changes.[1] The loss of India as a signatory, and, with this loss, the possibility that India will go on to manufacture bombs, is not trivial; it may well upset the halting of proliferation in other areas. If proliferation is an abstract issue, Indian bombs should be as upsetting as German or Belgian bombs. Yet the United States and the Soviet Union hardly seem ready to deploy the resources required to persuade India to sign the NPT; it is even possible that such half-hearted pressures for signature of the treaty have accelerated an Indian decision to produce bombs. Will India now abide by a treaty she intends never to sign? How long will India refrain from manufacturing a nuclear explosive? The list of arguments against an Indian atomic bomb is still long, but the weight of each item on the list has come into question.

THE COSTS OF A BOMB

Perhaps as significant as any consideration must be the reactions of the United States and the Soviet Union, the two superpowers

[1] A perceptive account of Indian policy on nonproliferation is presented in Shelton L. Williams, *The U.S., India, and the Bomb* (Baltimore: The Johns Hopkins Press, 1969). See G. G. Mirchandani, *India's Nuclear Dilemma* (New Delhi: Popular Book Services, 1968), for an Indian view of these questions; also see R. Rama Rao, "The Non-Proliferation Treaty," *The Institute for Defence Studies and Analyses Journal* I (July 1968), pp. 12-29, and Ashok Kapur, "Peace and Power in India's Nuclear Policy," *Asian Survey* X (September 1970), pp. 779-88.

that wrote the NPT and so unsuccessfully urged India to sign it. The array of carrots and sticks at the disposal of these powers is impressive, but not as impressive as in the past. What is more, such carrots and sticks must be coordinated so that India indeed receives benefits when she accommodates the superpowers and loses them when she does not; since 1968 no such coordination has been in sight.

To pressure India to sign the NPT, the United States presumably had some leverage in its policies on foreign aid, in direct programs and indirectly, as in World Bank decisions on non-project loan assistance. The Soviet Union had similar options in economic aid and military assistance programs and in its yearly U.N. Security Council votes on Kashmir. All such levers could have been used to pressure India to sign the Nuclear Non-Proliferation Treaty, but they were not. Nineteen sixty-eight saw the United States Congress reduce foreign aid and the Soviet Union offer military equipment to India's worst enemy, Pakistan.

As an alternative to quid pro quo inducements for India to sign the NPT, the great powers might have tried somehow to deploy enough military resources into the South Asian area to make nuclear weapons seem superfluous for India's defense against China. Yet such prospects were even more unrealistic than any of the levers suggested above because of the inherent difficulties in Soviet-American cooperation there and because of America's mood, which after Vietnam seeks fewer commitments rather than more.

Many Soviet and American tendencies thus conflict with the prerequisites for Indian signature of the NPT. Yet the NPT is drafted clearly to suggest Indian adherence; Indian nonsignature sets a precedent for other states to refuse to sign. If India "violates" the treaty in the 1970s by testing a "nuclear explosive" or bomb, the great powers' bluff will have been doubly called.

And what about contingency planning for such an obvious Indian affront? Perhaps the Soviet Union and the United States together would retaliate against India in a controlled and coordinated fashion, forcing New Delhi to dismantle any bomb stockpile she has created. Yet the superpowers' fears of a double-cross would be overwhelming, since either could so easily win Indian good will by being the first to forgive New Delhi for her entry into the nuclear club. It is rather more likely that neither power will lay any plans for serious retaliation but will rather seek to prevent a domino effect, after an Indian detonation, in Israel, Germany, and Japan.

There are many other countries that presumably will be affronted by an Indian move toward nuclear weaponry, for the outside world does not consist only of superpowers. India's image of pacifism would be seriously damaged, the respect of many aligned and nonaligned nations lost. But such arguments against bombs or for the NPT have also lost much of their punch within India, ever since the Chinese incursions across the Himalayan frontier in 1962. Many Indians since then have concluded that although the world professed to respect pacifism it really respected strength. Recent public opinion polls show a dramatic drop in the esteem for Gandhi's principles (albeit not for Gandhi the man); Indians have tired (rationally or irrationally) of being an object of other nations' sympathy, of visibly "needing help."

World opinion aside, India might at least show concern for the example she will set by moving to make a bomb or indeed by resolutely refusing to sign the NPT. Surely it was easier for South Africa and West Germany to dig in their heels against signing when they could be certain that India would always be out in the cold with them, and surely Indian procurement of a bomb or explosive would make it easier for Israel or Brazil to defy the superpowers in a similar fashion. Yet Indians express a monumental indifference. When India becomes the sixth nuclear power, she might begin to be concerned lest there be a seventh; until then, more immediate considerations outweigh any longer-term bogey the superpowers conjure up. America tells India to inhibit herself lest weapons spread to the Middle East; the Soviet Union similarly warns of proliferation to West Germany.

A more direct argument against Indian nuclear weapons alluded to Pakistan. It was definite that Pakistan would not sign the NPT unless India signed.[2] If India moves to exercise its nuclear options, surely Pakistan will be desperately driven to seek the same. No one in India credits Pakistan with an indigenous nuclear option in the remotest future; if the only possible supplier of such weapons is

[2] The official Pakistani statement of this stand can be found in United States Arms Control and Disarmament Agency, *Documents on Disarmament, 1968* (Washington, D.C.: U.S. Government Printing Office, 1969), pp. 317-24 (hereinafter cited as *Documents on Disarmament, 19—*).

China, it would be a great oversimplification to credit Indian moves with determining Peking's policies on further proliferation.

Although foreign economic retaliation in general will not definitely follow a bomb decision, Indians might still be averse to losing technical assistance, specifically in the nuclear field. The NPT certainly will require signatories to withhold any material assistance that might be used for weapons, indeed any assistance not subject to IAEA safeguards ensuring against bomb production. If other forms of outside assistance can not be orchestrated to encourage India to abstain from bomb production, perhaps the very treaty that India has refused to sign can be a tool for coordinating the sources of nuclear technology in this direction.

A question remains, however, as to how dependent India will continue to be on foreign expertise in this area and how effectively the NPT can function as an instrument of international embargo. If Indian authorities had really felt so dependent on outside help, their rejection of the NPT might have been less clear and less abrupt. The Indian nuclear program has come a long way since the 1950s; although Canadian and other outside assistance was essential before, it does not follow that it is essential to further progress. The Indian desire for a sense of independence is illustrated in plans to construct a fourth large power reactor at Kalpakkam in Madras, this time, the Indian government hopes, without having the condition imposed from outside that its use be exclusively peaceful. India can already produce a large percentage of the components for the Kalpakkam reactor, and Kalpakkam in turn could produce a significant quantity of plutonium, perhaps enough for forty bombs a year. The question now is whether foreign suppliers can be persuaded to deliver the remaining few components needed without asking for IAEA safeguards.

A straightforward interpretation of the NPT might require all suppliers to withhold such equipment; yet the nuclear field is rapidly becoming a buyer's rather than a seller's market with Canada, the United States, Sweden, West Germany, France, and other states all potentially underbidding each other. Thus India may yet get all she needs without signing the NPT, without accepting safeguards or offering assurances, without abstaining from the diversion of plutonium into explosives, peaceful or military.

Even if she does not sign the NPT, India might still decide to accept international safeguards on some purely peaceful projects, while rejecting them for facilities devoted to explosive devices. Uranium supplied under IAEA safeguards for a power reactor might thus free indigenously produced fuel for plutonium. It is not clear that the NPT would forbid this; certainly the likelihood of a liberal interpretation of the treaty will grow if the world does not react too severely against India's independence and/or if the desire for markets continues.

ECONOMIC COSTS

When all predictions about human behavior become uncertain or unconvincing, one usually turns in the end to the sounder, more material realm of economics. Throughout India in 1968 the ever reliable bulwark for those opposing bomb options was the argument that it was too costly, that the country simply could not afford to allocate large sums to a venture with so little practical return. If those who hold responsible positions within the Indian government continue to share this opinion, if those who are competent in accounting and systems analysis remain level-headed, then the world can perhaps relax its alarmed speculation about further proliferation, for India will be too sensible to exercise the option it refused to sign away in 1968.

Yet the ramparts of sensible economic calculation are not anymore so secure as a barrier to all forms of Indian nuclear weapons. Already a number of very reputable Indian economists have begun telling the public that plutonium bombs will soon no longer be expensive. If their calculations were oversimplified for 1970, they may not be so for 1975 or perhaps even 1973. As India very sensibly now begins a large-scale exploitation of nuclear energy for the production of electricity, supplies of plutonium will come into being. The separation plant for the enrichment of such plutonium to weapons-grade fissionable material already exists. It is thus more and more difficult to claim that a nuclear explosive the size of that used in Nagasaki will, by itself, be very costly for India in the year 1972 or the year 1975. Sensible analyses of the question within India quite rightly point to the costs of delivery systems capable of reaching the urban centers of China, of testing facilities for nuclear devices, of the uranium required for H-bombs, and of all the other accessories that the five nuclear-weapons states have felt to be necessary. No one yet has bought himself a big firecracker and been able to let it go at that; perhaps India similarly would not be able to stop with a cheap membership in the nuclear club. Yet if an Indian were to advocate only such a cheap entry per se, it would be less easy for critics to claim that it was not indeed cheap.

India's yearly defense budget has run in excess of one billion U.S. dollars; the current budget of the Indian Atomic Energy Department comes to almost 100 million. The 1968 Report of the U.N. Secretary General generally estimated the cost of a moderate program, one that would produce ten Nagasaki-sized bombs yearly, at less than two million dollars per warhead. It has been suggested that even this estimate exaggerates costs for an Indian project, now that streams of plutonium are circulating within the system and there is a separation plant in operation. Cost estimates for an operational, latest-generation missile delivery system are another matter, of course, but there are individuals in India who seriously wish to stop far short of such missiles.[3]

THE USES OF A BOMB

Although the strength of the arguments against nuclear explosives currently circulating within India is diminishing, this of course does not mean that the issue is decided; for a positive case for the bomb would still have to be discovered. Here again one can put together a list that is impressive in length but which must be weighed item by item.

There never has been a weapon for which a use could not be found. As many Western as Indian strategic writers have set themselves to composing scenarios in which an Indian bomb would be of great military value. Memories of 1962 conjure up hordes of Chinese Communist troops crossing the Himalayas; perhaps "tactical" nuclear weapons could be used to repulse them.

[3] A persuasive exposition of this view can be found in *A Strategy for India for a Credible Posture Against a Nuclear Adversary* (New Delhi: The Institute for Defence Studies and Analyses, 1968).

If the development of the small warheads for tactical situations were too costly (there is a great difference between a crude Nagasaki-type plutonium bomb and the warhead of a modern nuclear artillery shell), the cruder devices could still perhaps be planted as nuclear land mines in key valleys on the invasion routes, to be detonated only if an aggressor's forces filled the valleys.

Yet memories of 1962 are misleading. The Indian Army was unprepared for the war and was not nearly as strong in conventional terms as it is now. Furthermore, even in 1962 the Chinese chose to withdraw, in part because they would soon have been forced to. The logistic situation south of the Himalayas heavily favors India, just as the area to the north favors China; transporting supplies across the mountains is not easy. By "voluntarily withdrawing" in 1962, the Chinese may have caused India and other countries to forget how defendable India was even then, by conventional arms alone. In short, a Western argument today might be that India requires neither additional conventional forces nor nuclear forces to defend its northern frontiers but that she can make do with what she already has.

In terms of capabilities, China may thus not be so threatening.[4] Even in terms of intentions, one should not have accepted at face value Indian (or American) pictures of Peking as a totally unreasonable political force. Knowledgeable observers in both countries agreed that China's foreign policy has not been wild and irrational.

Yet hypothetical wars have a reality of their own, and Indians may still wish to seem stronger in case of any future outbreak. The Chinese might one day explicitly threaten to use their own nuclear weapons against Indian armed forces or Indian cities. In such a case perhaps Indian bombs would be valuable, if only to counter this threat by making it clear from the start that any Himalayan war would have to remain conventional because both sides could make it nuclear. Yet to be able to threaten Chinese cities may be a somewhat more costly proposition for India than simply assembling some plutonium bombs. China has air defenses; the aircraft currently available to the Indian Air Force or to Air India

---

[4] Assessments of necessary Indian responses to the threat of China are argued in Dilip Mukerjee, "India's Defence Perspectives," *Survival* XI (January 1969), pp. 2-8, and in K. Subrahmanyam, *The Asian Balance of Power in the Seventies: An Indian View* (New Delhi: The Institute for Defence Studies and Analyses, 1968).

International can not be counted upon to reach Shanghai or Peking. Missiles and missile warheads cost a great deal of money, and no Indian can deny that this might be the price of full-fledged membership in the nuclear club. One can easily enough imagine India entering the club halfway with a simple plutonium bomb and then feeling desperately obligated to tidy up its membership, its "retaliatory second-strike," with missiles and H-bombs.

Let us not however understate the military case for cheap plutonium bombs. No antiaircraft system is perfect. If France's entire *force de frappe* (some forty aircraft) is launched against Soviet cities, it is statistically quite likely that at least one Russian city will be destroyed. So similarly any Indian bomb stockpile, however crude the bombs and delivery systems, will force Chinese planners at least to contemplate a possibility that previously could be totally excluded, the possibility of losing a city to Indian retaliation.

When the Nuclear Non-Proliferation Treaty was presented in 1968, spokesmen for the superpowers might have argued that China was already deterred from using nuclear weapons against India by the likely responses of the United States and, later, of the Soviet Union. Indeed, as part of their effort to sell the NPT, the two states said they would have immediate recourse to the procedures of the U.N. Security Council after any such use of nuclear weapons against a have-not state which signed the NPT. For that matter, Communist China has over and over again sworn never to use nuclear weapons in combat unless someone else used them first.

It is likely that the possibility of a severe American response deterred any Chinese use of nuclear weapons against Delhi in 1968 and 1969, as well as against Tokyo, Bangkok, or Sydney. Soviet responses have probably contributed to deterring China in the 1970s. Even for a conventional attack, American and Russian assistance to India was quite likely. Indeed, Indian defense plans in 1968 could be rationalized as assuming a great probability for such moral and material support from the outside world in the event of a Chinese attack.

Only the more polemical Indian statements denied this; yet only the more polemical Western statements could claim that this summed up the issue. The existence of even a rudimentary Indian bomb stockpile would inevitably change the expectations of both sides about any new confrontation in the Himalayas, and the balance might yet favor India. Since the introduction of the support of the superpowers' nuclear arsenals on behalf of India was not plausible and salient enough in the minds of all the public that matter, Indian nuclear warheads would not be redundant. Any confrontation along a hostile frontier is sort of a game of "chicken," a test of whose resolve will come into question first. A visible addition of weapons capable of mass destruction to one side would not be without effect.

If India had never had occasion to inquire as to the specific extent of American or Soviet support, such support might have always seemed relatively obvious. Proposals such as the NPT, however, induced attempts at explication which could be quite upsetting. Hence, when the Indian government in 1967 sent high-level missions to sound out Moscow and Washington on the nature of future guarantees, both superpowers were far from definitive in the assurances offered. Since India's most likely enemy was still China, Moscow could hardly go as far as New Delhi would have desired; and the United States, even under President Johnson, was also intent on "building bridges" to China and on avoiding new exacerbations of the allegedly basic Sino-American conflict. It was hardly the moment for a very explicit guarantee to India. An American aversion to new commitments had certainly also emerged in the wake of the Vietnam war; and Secretary of State Rusk's testimony to the U.S. Senate, on behalf of the NPT, already circumscribing the implicit American commitments did not make the treaty any more attractive for New Delhi.[5] The joint U.S.-U.S.S.R.-U.K. statement suggesting immediate recourse to the processes of the U.N. Security Council, however intended, served similarly to convince many Indians that great-power guarantees were somewhat deceptive;[6] Security Council action could indeed be vetoed by France and by China (Nationalist China, and later—perhaps inevitably—Communist

---

[5] See Rusk's testimony in U.S. Senate, Ninetieth Congress, Second Session, Foreign Relations Committee, *Hearings on the Non-Proliferation Treaty* (Washington, D.C.: U.S. Government Printing Office, 1968).

[6] The American-British-Soviet Resolution on recourse to Security Council procedures can be found in *Documents on Disarmament, 1968*, p. 444.

China). Recourse to the Security Council in the event of aggression was called for in any event by the terms of the U.N. Charter; now the superpowers were suggesting that it would come only on behalf of non-nuclear signers of the NPT—a shrinkage of guarantees, perhaps, rather than an extension.

But it is indeed possible that specifically military uses of nuclear weapons are not crucial to an Indian decision to acquire them, any more than for the British decision to retain them. Returning to the sphere of economics, it was noted that purely peaceful technology would bring India quite close to bomb capabilities soon enough. The technology of peace and war is not easily separable; and one can even speculate that an Indian decision to "avoid" the bomb would be an obstacle to full research in the peaceful sector. Fissionable material will accumulate naturally over time, as will the knowledge of how to make it explode. Indeed some of the reluctance to sign the NPT, both within India and without, comes from the fear that surrendered options plus international inspection will be inhibiting to scientists.

More concretely, it is impossible to dismiss the possibility that peaceful nuclear explosives will be useful for liberating natural fuel resources deep underground; indeed, the U.S. AEC is an important source of optimism on this point. It is true that the NPT offers countries like India the use of peaceful explosives produced and controlled by one of the nuclear-weapons states. Yet we are trying here to guess whether India will make her own explosives, given the fact that no effective pressure seems at hand to make her accept those of the NPT system. For the moment, on scientific grounds, the guess must be "maybe."

One can also conjure up other economic benefits that an Indian bomb program might have, the material and psychological "spillover" effects it would have in perking up the Indian economic and scientific community. As with the U.S. space program, such arguments are inherently difficult to evaluate because of all the dangers of double-counting and misaccounting they present. Indian physicists working for the Indian Atomic Energy Agency program are for the moment busy enough getting the big new electrical power programs into operation. Five years from now something else might be needed to keep them busy and at home.

PRESTIGE

Yet again the Indian decision on the bomb can not be based only on material considerations of defense and economic cost. Political prestige is an important consideration in any nuclear-weapons debate, and only the most disingenuous outsiders would deny it. Few Americans or Europeans today know that India is capable of making nuclear weapons. Educated persons should be aware of the scientific prowess India has shown but they are not. It is useless to deny that the explosion of a rudimentary Indian bomb or peaceful explosive would make editorial writers and citizens all around the world sit up and take notice. The reaction of Afro-Asian countries to the Chinese explosion of 1964 was not to condemn but to show respect. Already there have been suggestions that India be offered some substitute form of prestige, such as permanent membership on the U.N. Security Council, to remove her psychological need for the bomb. Yet this all comes too late. Clearly, if India had not moved this close to a bomb, such suggestions would never have been made. One can conversely ask how much importance the outside world would have attached to Communist China if she had not entered the nuclear club. Would we not have passed Peking off during the Great Cultural Revolution as an internationally insignificant conglomeration of feuding factions, its economic house not in order, greatly overrated as an international actor? Bombs do make a difference.

It is easy to condemn any tendency within India to let something as "childish" as prestige decide the question of weapons of mass destruction; yet at least two of the current members of the club have defended their own programs in these terms. For the elite that makes decisions in India, the respect of a London taxi-driver or an American ambassador will count for something. It would be a better world if that respect were felt for those who had not made weapons ahead of those who had.

The question of relations with China brings out the interplay of these arguments quite interestingly. There are individuals within India who have been advocating new approaches to China because of their expectation that Peking will become more rational and reasonable now that the Great Cultural Revolution has run its course. Yet such persons also tend to be strong advocates of an Indian bomb program, on the grounds that successful negotiations with a reasonable China must be negotiations between near equals and that at least a nominal nuclear capacity will be required to endow India psychologically with this equality. Any Chinese reversion to greater bellicosity, for example with border incidents or testfiring a missile across India into the Bay of Bengal, would also increase Indian pressures for a bomb from a different camp, perhaps for a more costly and serious military program. Hence the current Chinese style leaves India moderately drifting toward nuclear weapons; any change in Peking's demeanor, toward or away from moderation, accelerates this drift.

Peking's orbiting of an earth satellite in 1970 was thus interpreted in India almost as negatively as an MRBM test into the Bay of Bengal would have been.[7] Knowledgeable observers could note that orbiting a satellite is much easier than developing an ICBM, since a missile requires a heat shield for reentry into the atmosphere and a much more accurate guidance system to locate its target. Yet many Indians (just like many Americans in their reaction to Sputnik in 1957) tended to view the Chinese satellite as a demonstration of ICBM capacity, or certainly of IRBM capacity, sufficient to deliver nuclear warheads to New Delhi and Bombay.

As noted above, superpower postures, even as of 1970, had left India somewhat unimpressed by the advantages of renouncing nuclear weapons and accepting the NPT. Two major changes in 1971 pushed India all the more toward a nuclear-weapons program. There was a dramatic rapprochement between the United States and Communist China at the same time that Peking won the seat in the United Nations. No one can predict how far this rapprochement will go in the wake of President Nixon's visit to Peking; yet in the past, speculation about nuclear threats involving South Asia would always have expected American strategic forces (perhaps joined by the Russians) to threaten China and to shield India. Such speculation already has had to be altered, with the United States remaining disengaged in its general withdrawal from the Far East or possibly, as part of its new

---

[7] For an Indian discussion of bomb options in the aftermath of Chinese satellite launching, see *The Institute for Defence Studies and Analyses Journal* III (July 1970).

alignment, even warning Russia not to attack China. This assuredly is the kind of change that can affect how decisions on nuclear weapons are made in India.

The second major event of 1971 to push India toward nuclear weapons was of course the secession in East Bengal that led to the Pakistani Army's brutal campaign of repression and then saw the Indian Army conquer all of East Pakistan, which was then proclaimed as the independent state of Bangladesh. Even before the Indian conquest, the internal breakup of Pakistan might have been seen as easing India's defense problems, as meaningful threats could no longer be directed at Kashmir. Yet the Indian government was motivated, by the plight of the refugees from Bengal or by the opportunity to defeat Pakistan more decisively or both, to do more than sit back on the defensive. The campaign into East Pakistan saw India draw closer to the Soviet Union in a Treaty of Friendship, while subjected to vocal Chinese attacks and to a thinly disguised American hostility. In the final stages of the Bengal hostilities, an American aircraft carrier task force even left Vietnam and entered the Bay of Bengal, quite obviously to deter India from exploiting its military victory over Pakistan to the very fullest.

Because of New Delhi's gratitude to Russia for the Treaty of Friendship, for the arms shipments, and for the timely vetoes of anti-Indian resolutions in the U.N. Security Council, perhaps Russia will be able to influence New Delhi; and perhaps this influence will be channeled to persuading India not to make atomic bombs. Since the splitting of Pakistan has eliminated what was the major military threat to India (a Pakistani attack on two fronts, intended to liberate Kashmir, supported by China), India has less of a real military need for nuclear weapons.

Yet considerations of military need will not by themselves settle Indian decisions on nuclear weapons. The experience of actual combat operations, and more particularly the recent specter of veiled threats by two nuclear-weapons states, can well have set India up emotionally to go ahead with the bomb. If Russian aid were too decisive and effective, India might not need it as much anymore and Moscow's antiproliferation leverage might not amount to much. If the Nixon administration's gestures in support of Pakistan seemed too demeaning for Indian prestige, New Delhi's bomb appetite has been whetted.

In a decision on military weapons, one would normally expect the military establishment of a nation to be quite significant. Happily for the tradition of democratic civilian rule in India, these opinions are far from decisive and not very openly expressed. Few generals in any army or air force will deny that they "need" any weapon or that they could profitably use it if it came to them as a free gift. It will thus always be possible for bomb advocates within the Indian Parliament to assure themselves that the officers of the IAF or Indian Army "would like" to have the bomb and that the morale of the average Indian foot soldier in the Himalayas would rise if he knew he had nuclear weapons behind him. Yet there has been little evidence of any enthusiasm for nuclear weapons in the Indian Army, or even in the Indian Air Force, simply because of the fear that a nuclear-weapons program would mushroom into something very costly, drawing funds from conventional weapons, which for the moment seem more urgent.

Like military officers everywhere, those of India thus can not be counted upon to oppose nuclear weapons on "arms-control" grounds. If a very limited and inexpensive bomb program comes along, their inclination will be to find requirements for it. Only while the budgetary predictions remain so open-ended will the military remain a barrier to India's exercise of her option.

THE COMING DECISION

If the styles of Indian reasoning are generally as outlined above, the question of an Indian bomb may no longer be simply whether to have one, but when and what kind to have. It may thus be valuable to summarize some of the momentum imported by events of the mid 1960s.

It is difficult to overrate the impact of China's military success in 1962 and the misleading impressions of Chinese strength and Indian weakness that it left. Not only Chinese behavior but the failure of Afro-Asian states to support India did much to make Gandhi's philosophies seem irrelevant to international relations. The nuclear question was most directly introduced then with China's entry into the nuclear club in 1964. Demands within the Indian Parliament for a matching nuclear-weapons program emerged strongly in that year and increased, into 1965 and 1966, with the successive Chinese tests leading up to an unexpectedly early H-bomb. Nineteen sixty-seven, the year of serious drought and threat of famine in India, saw discussion of nuclear-weapons options recede somewhat in the Indian press, but the emergence of a Russian-American draft of an NPT in late 1967 and 1968 seriously enhanced Indian interest in the retention and possible exercise of the weapons option.

For a time serious pressure from the United States and the Soviet Union for an Indian signature was anticipated; responsible officials in New Delhi girded themselves against this pressure, while a few suggested that India prepare to capitulate and sign. The pressures did not materialize, but the effect of the anticipation was nonetheless important. As in other countries, it can not easily be denied that the NPT has had the effect of whetting many appetites for bombs. Where the treaty will be signed and ratified, it may serve its intended purposes; where it has been considered and rejected, however, it may well have been counterproductive.

An Indian refusal to sign the NPT may thus not yet have been absolutely determined early in 1968. Mrs. Gandhi's decision might have been signaled in a statement saying that India could not sign the treaty "under present circumstances." If this did not signal a decision, the response of India's galleries of political opinion was so enthusiastic as to make it one, as the focus shifted from the conditional to the negative nature of the statement.

India's press had by now largely arrayed itself against the treaty. One might have suspected that the better newspapers would still be managed by members of the same "old boy network" that dominates the upper ranks of the Indian civil service—moderate in tone, attuned to foreign thinking and opinion. Yet the editorials of the major newspapers are also increasingly coming under the influence of younger, more nationalistic writers, and the impact of these newspapers in India is quite significant.

Reporters on the New Delhi scene had noted a surprising increase in the use of great-power military strategic jargon in 1968 in conversational situations of the casual cocktail party variety. Phrases such as "assured second-strike capability" or "credible first-strike scenario" are only superficial symptoms of where the Indian elite's imagination may be turning. Conjoined with this is the great popularity of the power-political abstractions of Pierre

Gallois. For those new to the analysis of strategic weapons, the apparent clarity of such logic generally has an initial attraction: note, for example, the assurance of many American writers in the late 1950s that the United States could never rationally go to World War III in defense of Europe. Late in 1968 one Indian government official after another could be found commenting in almost identical terms that "no nation ever helps another except out of its own selfish interest"; that is, the United States would never retaliate against anyone that dropped a nuclear explosive on New Delhi.

At the end of the 1960s the nuclear-weapons decision was still often discussed as a question of monetary cost. For as long as the responsible officials of the central government and the Congress Party thought of a nuclear arsenal as involving large expenditures of capital and foreign exchange, the demands of the opposition benches could yet be ignored. If polls of Indian public opinion now show some 70 percent of literate Indians to be in favor of an Indian bomb,[8] this does not indicate the response there would be if the price tag in inflation or higher taxes were clearly specified. India is a large country. Most large countries in the world find it difficult to coordinate their bureaucracies behind great new projects of substantial deviation from past policies. It is quite fashionable in India to indict the central government in New Delhi as uncoordinated and unmobilized, unprepared to carry a project through to its logical conclusions. However unfair this picture might be, however typical this pattern may be for all great bureaucracies, it suggests that a "crash program" would not be set in motion from the top to give India "a nuclear arsenal whatever the cost."

Thus the decision to produce bombs once might have been a difficult one to make in New Delhi. But what if the decision becomes much easier since the cost of a simple plutonium bomb has decreased? Here now the attitudes of Indian scientists and bureaucrats at lower levels may be crucial, for if an attractive enough prospectus is written for an underground peaceful nuclear explosive detonation in the 1970s, it is less certain that the political leadership will want to reject it. India, to repeat, is a large country. Large countries can undertake prestige projects if they

[8] *Public Opinion Survey* of the Indian Institute of Public Opinion, vol. XIII, no. 5 (February 1968).

want to, even if their per capita income is significantly less than that of smaller states. India has undertaken such a project in the development of electrical power production by atomic energy, the economic sensibility of which was questioned in the 1950s but is questioned no longer. Atomic energy is a field in which Indians rightly take pride; it is a field in which scientific morale is high.

Scientific morale is high, in part because Indian physicists have been exempted from some of the bureaucratic nuisances that too often show up in India, in part because they are engaged in some big projects that are showing clear progress. The two go together. When the big reactors are in operation in Rajasthan, when construction of the newer ones in Madras is well under way, it may seem important for the Indian atomic energy community to move on to something else that is big and productive and exempt from bureaucratic red tape. The nuclear explosive project clearly is one project that fits this description.

THE "PEACEFUL BOMB"

A project for a "peaceful" nuclear explosion (to avoid seeming bellicose) underground (to avoid violating the test-ban treaty which India did sign) with plutonium as the fissile material (to avoid great expenditures) thus may win whatever support it needs among both scientists and politicians. Among Indian public officials known to be opponents of nuclear weapons in 1968 almost none were prepared to rule out peaceful explosive devices along with explicit weapons. Some persons drawing this distinction may quite sincerely see a major demarcation line here; hope even attaches to the possibility that such devices would be physically different from weapons, although few scientists anywhere would reinforce such hope. Many other Indians know quite well however that these "peaceful" explosives are functionally usable as bombs and that every editorial writer from New York to Tokyo will immediately note this overlap; in truth, such Indians welcome the overlap, for it would allow India to join the club behind a euphemism avoiding too brazen an affront to the international community and behind a civilian economic rationale that may suffice to push the project through domestically.

There will be some credibility problems in any Indian underground "peaceful" explosion; if the possible lack of credibility abroad is taken into consideration in the Indian domestic decisionmaking process, some delay on a detonation might yet result. A Nagasaki-style plutonium bomb is not the ideal instrument for liberating natural gas or oil deposits; the ideal peaceful explosive would cause less radioactivity and would be more compact, so as to require a well of smaller diameter for its burial. It is not even clear that geologically promising areas for the application of a nuclear explosive will have been identified. Other countries, particularly the Soviet Union, have taken part in, or are privy to, a portion of Indian geological survey work and would presumably be able to question the honesty of any Indian claims on the purpose being served by a particular explosive test. The costs of a modest underground test program have been estimated in U Thant's report as about fifteen million U.S. dollars;[9] ninety-nine million U.S. dollars has already been allotted for Indian nuclear research.

Yet the credibility of India's immediate peaceful nuclear explosives projects may not be so powerful an argument for outsiders to raise, for the fine print of the Indian announcement could simply note that the detonation was an underground test of an Indian peaceful nuclear explosive which from time to time in the future would be utilized in the tapping of mineral resources. If France must test its weapons before dropping them on cities, why can India not test its peaceful explosives underground before immersing them in mineral deposits?

If economic costs are not excessive, if bureaucratic momentum is not too difficult to generate, the "peaceful bomb" can indeed offer dividends to a political incumbent. It has been said that public opinion on Indian nuclear weapons is not to be taken seriously, since few votes in India are ever won or lost on foreign policy questions. But this is only partially true. To mobilize the individual party workers necessary to keep itself viable as a national party, the Congress Party has always needed a national focus in order to distinguish itself from all that is communal and particularist and regional in India.

[9] *Report of the Secretary-General on the Effects of the Possible Use of Nuclear Weapons and the Security and Economic Implications for States of the Acquisition and Further Development of These Weapons* (New York: United Nations, 1968).

Some of the "national" interest of this Congress Party elite is founded on foreign policy, and it has included a growing aversion to outside aid, outside commitments, outside influence. To some extent these trends reflect a simple generation gap; government officials educated in the older British system almost inevitably show more admiration for the "reasonable" world consensus. If there was in India any sentiment at all in favor of signing the NPT, it was in the senior civil service ranks of the Ministries of Finance, External Affairs, and even Defense. At more junior levels of the Indian government no such protreaty sentiment appears to exist, while probomb sentiment is anything but lacking.

The 1969 split within the Congress Party thus had very interesting consequences for the likelihood of an Indian bomb program. As in every democratic country, much in India must depend on the calendar of elections. If the Congress Party had stayed united and parliamentary elections had been held as scheduled only in 1972, strong pressures might already have emerged for a 1971 token detonation of a plutonium warhead, simply to increase the electoral appeal of an incumbent administration. The split of the Congress Party left it quite unclear, however, as to who would have credit for such an Indian bomb. Bomb proponents and bomb opponents were to be found in Mrs. Gandhi's camp and among her Old Congress opponents, as the struggle hinged more on domestic and personal issues than on questions of foreign policy. Both the Old Congress and the New Congress could have claimed incumbent credit for any bomb that was clearly the fruit of the late 1960s. Mrs. Gandhi, in advancing the parliamentary elections to the spring of 1971, won a decisive victory, pushing back the requirement for the next electoral test to 1976. Inadvertantly this may have done a real service to the cause of the NPT all around the globe. If the elimination of immediate electoral considerations at all postpones the first Indian detonation, it gives the rest of the world more time to settle into commitment to, and compliance with, the NPT system. An Indian detonation in 1975 may be far less disruptive than one in 1972; it may be far less likely to stir public opinion in Japan or Italy to the point that those countries demand similar nuclear privileges merely because they are envious.

The margin of Mrs. Gandhi's victory might even suggest that her government will never need to contemplate nuclear explosions merely for domestic political appeal. With more than two-thirds of Parliament controlled by her New Congress Party, she has the legal authority to revise the Indian constitution whenever her social programs make it necessary. Yet great electoral victories produce not only great mandates but also great expectations. After the Indian princes have been divested of their incomes, after the administration of Bangladesh proves to be economically costly as well as militarily advantageous, it may be much more difficult for the New Congress to find dramatic ways of bettering the life of the average Indian or of increasing the stature of India as a nation. If India's domestic problems are basically close to intractable or at best only susceptible to the most gradual of solutions, a sense of disappointment with Mrs. Gandhi, however unfair and unreasonable, is likely to set in by the third and fourth years of this parliamentary term. For the reelection of a New Congress majority in 1976, therefore, a bomb detonation in 1975 could still be a help.

WORLD REACTIONS

The crossing of this Rubicon by 1975 or any other date of course still depends on some outside variables. It was suggested above—pessimistically—that *any* change in Chinese behavior is likely to accelerate the development of explosives. Much will also depend on what other states do about the NPT. Indians will generally feel that a decent interval should be allowed between rejection of the NPT and violation of it. If most significant states refuse to sign or to ratify, India may have a relatively greater freedom of action, as long as she is not considered to be responsible for other states' defiance of the superpowers. For a time in late 1967, it seemed that India might be trying to organize states such as West Germany, Japan, Sweden, and Italy into a nonsigning bloc, a bloc that swapped arguments against the treaty and regularly consulted at a high level on what stands to take, and so on. In the spring of 1968, perhaps at great-power behest, New Delhi apparently decided to avoid agitating against the treaty rather than simply to state its own unwillingness to sign. At the Conference of Non-Nuclear Weapons States in Geneva in the late summer of 1968, for example, India in fact voted with the pro-NPT states on a number of points. It is indeed possible that the failure of the superpowers to align many carrots and sticks behind the treaty reflected a resignation to India's abstention in an exchange whereby India at least ceased being a major propagandist against the treaty. India's goal might thus well have been for many others to join her in refusing to accept the NPT and for herself not to seem to be the cause of these other abstentions.

If widespread acceptance and ratification of the NPT is delayed, an embargo obstructing any early freedom from outside commitment will be less likely and India may sooner acquire plutonium supplies eligible for military use. Yet even this may not be so crucial, especially if the euphemism of "peaceful explosives" is to be used. India's commitment to Canada concerning the plutonium currently being used and stockpiled is vague, calling simply for "peaceful use." While Canada has ventured some interpretive comments to the effect that this would exclude peaceful explosives, no one in India has declared himself bound by such an exegesis and the Canadian government may be reluctant to press it very vigorously.

India's freedom of action on the nuclear question may not always depend on many nations abstaining from the NPT, and it is even paradoxically possible that a general acceptance of the treaty will carve a slot for India in the nuclear club, perhaps as the last member. If others sign the NPT and India does not, the world will soon come to see three categories of states: weapons-possessing, weapons-renouncing, and India. The treaty strictly compels signatories to the treaty not to cooperate with India if she moves toward weapons; but India could throw her civilian facilities open to IAEA inspection like any other weapons state, while keeping her "military" facility closed, with no real economic handicap. Most importantly, the risks of a major political reaction against India in the form of markedly worsened relations with the Soviet Union or the United States would be considerably lessened. India might simply be the one nation that never promised to forgo nuclear explosives and hence was not violating any promise. By doing all its testing underground, it would be keeping its promise of the test ban. If the NPT is based on fears of large or infinite numbers of nuclear states rather than the number six, the world might well yet adjust to Indian membership in the nuclear club.

The failure to apply serious pressure for Indian signature of the NPT indeed suggests a rather less frightening forecast of what Soviet or American reprisals would be after an Indian nuclear detonation.

While Indians typically claim to see no link between their own decision to produce a bomb and the decisions of a seventh, the professionals of the Foreign Ministry are privately quite aware of a real linkage here. To pretend to miss the linkage was prudential, simply to ward off what was part of Soviet and American pressure for signature of the NPT. If the Indian government does not want to be followed into the nuclear club by two or three additional new members, it could of course simply decide never to become a club member itself. If this seems too high a price, New Delhi can still at least give careful consideration to the timing of any first detonation, since the two superpowers may have their NPT house much more in order at one time than at another. If Japan and West Germany had already ratified the treaty, for example, and a discrete time had passed in which public opinion had shifted to other issues, an Indian detonation could have less of a domino effect. When other countries seem to be looking for excuses to reject the treaty, however, the Indian ambassadors would surely have to report home that an Indian bomb would have effects outside of India.

The two superpowers of course can pave the way for treating New Delhi as a special case instead of lumping India in with all the other potential "nth" countries, all equally to be urged not to become the sixth. If there is little discussion of Indian special circumstances, it may be harder to decouple a seventh from an Indian detonation, and this might yet deter New Delhi from going ahead. Yet at some point Indian patience might wear thin, at which time a message would probably be sent to Moscow and Washington to the effect that an Indian explosion would come soon regardless of the consequences for the NPT. If the United States and the Soviet Union did not then begin to redraw their line of containment on nuclear weapons spread, any domino effect would be their responsibility.

One can thus again pessimistically paint a narrow range of inhibition around an Indian decision for plutonium bombs. Either a total breakdown of the NPT system or an almost total success (all but Indian subscription) could free New Delhi to act as she pleases; somewhere in between lie situations more inhibiting to India, less conducive to the world's acceptance of India as number six.

Presumably India's neighbors might find it especially difficult to adjust to Indian membership in the nuclear club, particularly Pakistan and perhaps Communist China. It has been difficult to predict the Pakistani reaction to an actual Indian decision to manufacture weapons, for its official propaganda has already long accused India of having made this decision. If Rawalpindi has gained any international sympathy by such accusations, it simultaneously has muddied the waters of its own contingency and intelligence planning systems. That some serious fears were circulating is illustrated by Pakistani efforts to pressure India into signing the NPT, for example, suggesting the inclusion of great-power security guarantees in the treaty but *only* extending to signatory states.

The time may come when a rump Pakistan will also be able to manufacture plutonium bombs—a fact that Indians must yet consider—but that time is far into the future. For the moment a frantic effort might be made by Pakistan to acquire (or at least to give the appearance of acquiring) such weapons from China, but a favorable response to such requests is also not likely. Alternatively, a Pakistani regime might try to assume a "holier than thou" posture, renouncing nuclear weapons and focusing world condemnation on the state that brought such weapons into South Asia. If world condemnation is as insubstantial as has been assumed here, this latter response is only optimum by default.

It is indeed true that non-weapons states have stood up well against weapons states in several cases, such as Egypt versus Great Britain in 1956, Hanoi versus the United States since 1965; possession of nuclear weapons perhaps inhibits a nation's confrontation with an opponent who lacks them. Yet Hiroshima and Nagasaki in 1945 obviously demonstrate that this does not always hold true, and there are circumstances in which Pakistan might suffer heavily because India had acquired nuclear means of inflicting such suffering.

The response of China has been at least as difficult to predict. There was a time when Peking seemed to be defending or encouraging the acquisition of nuclear weapons by all states. Significant Chinese addenda suggested that it was better if each state developed its own weapons, rather than seek foreign (that is, Chinese) assistance; yet Western observers still feared that China might give such weapons away, or certainly would not oppose their spread. Peking has consistently denounced the Nuclear Non-Proliferation Treaty as a Soviet-American plot. Of late, however, oblique Chinese references have advised that it would be inappropriate for Japan or India to acquire nuclear weapons, suggesting that Peking may fear lest its earlier propaganda be taken too seriously. The "pro-Chinese" Communists in India have taken the stand that India does not need nuclear weapons since she has nothing to fear from China.

Hence it would seem that China has not eagerly looked forward to an Indian nuclear detonation, and that Pakistan certainly has not. Yet the responsive steps either would take are hardly clear enough to be a major dissuasive element in the Indian decision. The United States and other countries can attempt to frighten New Delhi with visions of Chinese preemptive attacks on unprotected Indian plutonium facilities or airbases or first-generation missile sites, but such a preemptive war is no more likely than were American attacks on China in 1964.

The general consequences of this line of antibomb argument may be all the more unfortunate. Military analysts both in and out of India have noted that plutonium bombs are not first-rate weapons anymore, that they are perhaps even meaningless as weapons. The latter is indeed an exaggeration, but Indians may yet feel driven to spend large sums of money on later-generation warheads and delivery systems, having become addicted by the token dose of plutonium bombs. In effect, it has happened this way five times before. Why not in India? There are indeed people in India demanding an immediate full-scale military program with "hardened" second-strike missiles capable of reaching Peking and hydrogen warheads to match those of China. India's adoption of such a program in lieu of the peaceful plutonium explosives approach is unlikely because of cost and because of affront to the outside world.

Analysts who see the biggest and best as a necessary long-term goal for India can either favor or oppose an interim "cheap" plutonium bomb program. Some Indians have argued that only the token bomb will suffice to mobilize the country for the full-fledged military undertaking, which exactly parallels foreign warnings that India would not remain content with a cheap program. Other analysts, including some who are very influential within the Ministry of Defense, are averse to token plutonium projects; they fear that this would produce premature crash programs where more orderly nuclear-weapons development schemes are required or even suggest preemptive attack from China while it alone has an up-to-date nuclear arsenal. In terms of India's image, an early detonation without a rapid follow-up to H-bombs and advanced delivery systems might also make India look foolish and incompetent for the longer run, as Indian postdetonation progress will inevitably be compared with that of Communist China. In terms of prestige, a great number of foreigners would of course be impressed by any Indian nuclear detonation; yet more knowledgeable observers might then be negatively impressed, after a time, when they realized how much of a token gesture the plutonium detonation had been. For each year in which India forgoes a token detonation, for whatever reason, the infrastructure of technology and industry required for a respectable follow-up weapons program will have increased.

In effect, there are four positions on nuclear weapons within India: (1) one made up of those who are content with a "cheap explosion," (2) one quite unrealistically favoring an immediate crash program for the biggest and best weapons, (3) one favoring the "cheap explosion" as a catalyst to induce a longer-term program, and (4) one favoring a more measured approach to the biggest and best. The first group may well have a strong enough case to break the ice and defy the NPT. The third group may then however get its way, perhaps at costs very difficult for India to bear.

Perhaps the current nuclear powers could still offer one very strong argument against India's beginning even a token explosives program, namely, the example of their own inability to be content with such a program. It is psychologically difficult for the nuclear-weapons states to phrase arguments along these lines; moreover, India may be rational enough and poor enough to be content where others are not. If Great Britain is slowly backing toward a token nuclear capability and still derives a sense of benefit from it, India might yet be able to stick with the same.

Atomic weapons are capable of inflicting the most tangible suffering on large masses of mankind. One is thus tempted to seek very tangible assurances that one country or another will not be able or willing to procure such weapons. For the moment no such assurance exists in the case of India. An Indian signature of the NPT might have supplied that assurance, but this truly seems out of the question now. Economic costs have lost their decisive punch as an argument against bombs, or will soon enough; even those most responsibly concerned about costs today are only arguing for a gradual and carefully planned bomb program, not for total abstention. Factors as intangible as the international political mood or the domestic mood in New Delhi may yet suffice to postpone or preclude an Indian bomb decision. Yet at the moment this postponement hardly promises to be forever.

5

Jerusalem

Among the significant near-nuclear states, Israel since 1968 has been remarkable in terms of the extent of its "conventional" military problems, which for a time distracted Israeli citizens from the pros and cons of signing the NPT but which can make nuclear weapons increasingly attractive as other "solutions" seem to fail. The conquests of the 1967 Six-Day War resulted in boundaries that could more easily be defended against conventional attack but perhaps were less secure against random terrorism. No tendency towards recognition of Israel emerged in any of the Arab states, as the obvious loss of territory left Jordan, Egypt, and Syria seemingly incapable of controlling their extremists or of contemplating compromise peace proposals. Soviet moral and material support for the Arab states did not decrease, while a French embargo forced Israel to turn to the United States as its major arms supplier. The atmosphere thus precluded many of the supposed advantages of the NPT. A mutual renunciation of nuclear weapons could not induce waves of political good feeling that would multiply out into real détente in the Middle East; distrust was already at such high levels that either side would from the first assume no good faith from the other. Israel might thus find enough advantages in withholding signature to outweigh any marginal losses.

REJECTION OF THE TREATY

For a time it was thought that the United States might possibly come to want the NPT enough to be willing to offer substantial assistance and support in exchange for Israel's signature. At the time of the long delay on the first American sale of fifty Phantom jet fighter-bombers to Israel, it was rumored that Israeli signature of the NPT could have accelerated the American arms transfer; in the event, the aircraft were promised without Israeli acceptance of the NPT. Russian involvement on the side of Egypt and the increased activity of Palestinian guerrilla groups have indeed involved the United States, however reluctantly, in continual military support of Israel. Although the Nixon Administration might have wanted to be more neutral than its predecessors or to apply more pressure for things like the NPT, Soviet reinforcement of the Arab side has nevertheless tended to narrow its options here.

There were also rumors in 1968 that Israel was demanding the reestablishment of diplomatic relations with the Soviet Union, which was one of the depositories for the NPT, as the price of signature. The reappearance of a Soviet Embassy in Tel Aviv, after the rupture of relations during the 1967 war, would once again affirm Israel's right to exist and would take some steam out of more extreme Arab positions.

Aside from the possibility of such great-power concessions, many Israeli officials see a marginal value in staying out of the NPT system if it means that speculation about Israel's nuclear potential would psychologically intimidate the Arab states. The theories of strategic psychology propounded here are quite different from those used by Americans to describe their balance of terror with the Soviet Union. Israelis are not so directly concerned with carrot-and-stick games of encouraging the Arabs to do this, deterring them from that. Signing the NPT in order to get Arab states to sign has no logical appeal; there is also little speculation that if Israel had nuclear weapons she could decisively deter Arab attacks. Rather than attributing a rational calculus of costs and gains to their Arab adversaries, Israeli planners impute a more visceral sense of power and weakness. By refusing to sign the

NPT while for the moment not making any bombs, Israelis do not guide the Arabs, but rather hope to cow them with a vaguer sense of power which can be furthered by periodic rumors of bomb projects. Arab propaganda inevitably assists in this purpose by claiming that Israel is already manufacturing nuclear weapons; if enough Arabs come to believe this, it might be gratuitous to dispel their illusions.

It is difficult to get Israelis to take seriously any of the reciprocal advantages of the NPT. The Arab states are rated as being incapable of manufacturing their own atomic bombs at any foreseeable time. This is perhaps a little too final, for the U.A.R. does have a *plan* for a 150-megawatt power reactor near Alexandria, which would yield sufficient plutonium for thirty atomic bombs a year; however, the project would require Soviet or other outside assistance, which is not yet assured.[1] The threat of such an Arab counterpopulation weapon is moreover already vitiated by the existence of certain functional equivalents. Israel is the one near-nuclear country whose cities are already subject to international violence. If Al Fatah can set off TNT bombs in Tel Aviv, who will worry about an Egyptian atomic bomb fifteen years from now? Moreover, Egypt's use of poison gas right through its intervention in Yemen, and rumors of plans to use it against Israel in the 1967 war, adds another counterpopulation weapon which might kill Israeli civilians at rates comparable to those of nuclear weapons.

There are additional Israeli objections to the NPT which may resemble those of other nonsigners. The imposition of safeguards is expected to be a nuisance for any nuclear research and energy program, as valuable time would be wasted in answering inspectors' questions and in proving that fissionable materials had not been diverted; the Israeli government, for reasons real or imagined, feels a special aversion to the IAEA. Because of the regional structure of the election process, it is almost impossible for Israel to get a seat on the IAEA Board of Governors; similarly, no Israeli nationals were employed as IAEA inspectors in 1968. If the procedures of the Vienna organization were thus biased, the inspection process could easily enough be used to make trouble. Voting in the U.N. General Assembly these days shows greater

[1] See Leonard Beaton, "Capabilities of Non-nuclear Powers," *A World of Nuclear Powers?*, ed. Alastair Buchan (Englewood Cliffs: Prentice-Hall, 1966), p. 21.

sympathy for the Arab states; if the nationals of these neutral European and Afro-Asian states were similarly biased when serving with the IAEA, they could generate unwonted charges of military activity or require unjustified shutdowns of certain facilities, merely as part of a policy of harassment. Commercial espionage is of course a possibility for the longer run. For the short term, military espionage would also be a great threat, as the tactical and strategic situations again become intermeshed. An inspection of the reactor at Dimona might ensure that Israel was not manufacturing plutonium bombs; yet details obtained in such an inspection might also aid an Arab air force planning a conventional reprisal attack with the intention of simply polluting part of Israel with radioactive debris. The deployment of some of Israel's American-supplied Hawk antiaircraft missiles around Dimona does not prove that the facility will be used to produce weapons; even under present circumstances it can be a target for Arab weapons.

Israel might be more willing to undergo inspection by American officials only, but the current treaty draft rules out such an arrangement. There has moreover been a general aversion even to American inspection of the reactor at Dimona; American visits to Dimona, when tolerated, are kept quiet and never attributed with the legal status of an inspection. The touchiness of the Israeli regime here reflects nationalistic objections that are voiced by the parliamentary opposition; but it also shows that the policy of "keeping the Arabs guessing" still holds. Americans probably try to reassure Cairo, on the basis of what they have seen, that Israel is not making a bomb. As long as explicit American or IAEA inspections are not allowed, however, Arab propaganda will probably muddy intelligence channels enough to preclude full acceptance of such assurances, which is exactly what the Israelis want.

Had the NPT been written without an inspection clause, an Israeli signature might indeed have been more likely; this may simply show the need for the safeguards of Article III even if the article proves to be a nuisance in other parts of the world. Israel and the Arab states have an unhappy history of violated agreements, and it is clearly unlikely that a pledge not accompanied by inspection would decisively preclude any state in the Middle East from making weapons. When one's national existence is at stake, treaties can be ignored, or they can be signed in a relatively carefree manner.

It can readily be seen that other arguments for the NPT have carried no weight at all in Israel. Israel's nonadherence to the treaty clearly makes it easier for other states not to sign; aside from most of the Arab states, which may implicitly make their ratification contingent on Israel's adherence to the treaty, countries with more significant nuclear establishments can count on not being out in the cold alone as long as Israel stays out. Yet no one in Israel really broods about Israel's contribution to India's retaining and someday exploiting a nuclear-weapons option. Similarly with Japan or Brazil. Even the "threat" of West Germany, which for historical reasons might easily rouse great concern within Israel, does not make the NPT attractive. The idea that West Germany is somehow counting on Israel to make its own nonadherence to the NPT less noticeable and unpalatable has not been widely circulated within Israel, and when circulated it has brought little response. Within more knowledgeable governmental circles, the connection was complicated in that a leading German opponent of the NPT is Franz Joseph Strauss, who was important in clandestinely channeling West German arms aid to Israel in the late 1950s and early 1960s—a "friend" therefore.

All the Arab states except Algeria and Saudi Arabia signed the NPT on the very day that it was open to signature, but ratifications have come quite slowly. Israel professes not to care. The Arab states could not explicitly make their signature or ratification conditional on Israel's, for to do so might hint at a recognition of a political entity by that name. It is nonetheless possible that the rush to sign the NPT reflected a sincere concern that Israel be pressured to sign, that Israeli nuclear weapons would otherwise come into existence. On the other hand, it may have reflected, more simply, a feeling that Israel probably would not sign and that Arab signature would therefore look comparatively good in the eyes of the world. Finally, the rush to sign may have reflected the Arab dependence on the arms aid of the Soviet Union, which now attaches high priority to the NPT.

ISRAELI NUCLEAR INDUSTRY

There is little doubt that Israel possesses the resources and expertise necessary for nuclear explosives; one can thus say (as one can say about ten or more other countries) that Israel is "making preparations" for the manufacture of atomic weapons. After some years of French assistance, the Israeli program includes a reactor at Dimona which can produce from five to seven megawatts of electrical power, or from five to seven kilograms of plutonium (about enough for one bomb) a year.[2] There apparently are no French safeguards to assure peaceful use of the materials produced. Now approaching self-sufficiency in nuclear technology, Israel is perhaps capable of undertaking the construction of larger power reactors and of separation plants for the reprocessing of plutonium, perhaps even more advanced projects. The assembly of fissionable materials into weapons also is not beyond Israeli competence, allowing a relatively small period yet to complete the basic research. It is only in the supply of uranium that Israel might possibly for some time remain dependent on outside sources, although enough such material might yet be extracted from phosphates to support the Dimona reactor.

It would be one thing for a country like Israel to postpone any explicit decision to manufacture and demonstrate its possession of nuclear weapons; it would be another to halt the scientific progress which inevitably draws the nation closer to weapons production. Several steps can be taken to bring such production to within months or weeks, steps which may soon be entirely justifiable in terms of civilian benefits only. Newer and larger reactors may come into use for the production of electrical power, incidentally producing larger flows of plutonium. Plutonium separation plants might then be advisable for the processing of such material for reuse as reactor fuel. Still further along, an investment in a uranium enrichment plant might become advisable to produce the basic fuel for reactors indigenously (either plutonium or enriched uranium can be adapted for use in nuclear explosives).

One might question whether Israel will really consume enough electrical power to justify large power reactors. The answer would be negative but for Israel's option of electrically desalinating sea water, which could make fresh water available to all the Middle East. If such projects are undertaken nationally, large quantities of plutonium capable of being converted into from fifty to one

[2] A concise description of the evolution of the Israeli nuclear program and the facility at Dimona can be found in Leonard Beaton, *Must the Bomb Spread?* (Hammondsworth, Middlesex: Penguin, 1966).

hundred bombs a year may be generated. For the moment American cooperation may be required in the research on how to use the electricity generated by reactors in the desalination process. A number of Israelis believe that the United States has deliberately been dragging its heels on such projects for fear of precisely the prospects listed above. Vague hopes for reactor-desalination projects jointly administered by Egypt and Israel hardly seem practical; however clearly they might solve the arms-control problem, they would first have to solve almost all of their political problems. Opinions on whether Israel really faces an imminent water or electrical power shortage differ. A year of bountiful rainfall can induce optimism, a drought the reverse. Americans suggest that Israelis are exaggerating their needs in a move toward earlier weapons options; Israelis suggest that Americans have been judging the area's agricultural needs in terms of arms control.

If Israel persists in rejecting the NPT, the United States will probably refuse to cooperate on the desalination projects. A claim will be made that the NPT binds the United States to such a refusal. Yet this would appear to be quite false, since the United States aid would not apply to nuclear reactor projects per se but rather to the use of electricity generated in such projects. The United States might be no more bound here than it is bound not to sell India light bulbs illuminated by electricity from Indian reactors operated in the absence of IAEA safeguards. It is also possible that Israel might need assistance in the construction of large reactors, which more clearly comes within the purview of the NPT. Alternatively, Israel may slowly undertake the indigenous development of both reactors and desalination plants, with the result that fissionable materials not safeguarded will still eventually appear in quantity.

Unless IAEA inspection is commenced, it will thus be difficult or impossible for the world to assure itself that Israel has not entered the nuclear-weapons club. Yet even if such inspection does commence, the task of the IAEA inspectorate will be more difficult in the Middle East than anywhere else. If the institution of IAEA safeguards were to remove American resistance on desalination projects, the quantities of plutonium circulating in Israel would still keep Arab propagandists buzzing about Israeli weapons. The NPT moreover does not rule out nationally-operated plutonium separation plants, and Israeli signature might reduce Western alarm about such a plant. If fissionable material were thus to become available, would Israel's military nuclear option really have been lost, or might it instead have been enhanced?

If Israel arrested and deported the IAEA representatives while seizing safeguarded material for bombs, the U.N. Security Council would immediately have to consider the act. But Israel has defied the U.N. before. So brazen a diversion of plutonium into bombs would perhaps only occur in the case of some crisis seriously threatening Israel's defense position. More clandestine diversions would also be possible, however, given the inherent uncertainties of IAEA inspection procedures. If a 2 percent range of error is inevitable, perhaps 1 percent of Israeli plutonium could be milked off for secret assembly into bombs, as reinsurance for an eventual day of need. Arab propagandists would continually be accusing the Israelis of this, as might some non-Arab IAEA inspectors friendly to the Arab cause; with this much noise in the system, the Israelis then (as today) might be able to keep a clandestine program hidden.

It is far from inevitable that Israeli adherence to the NPT would so paradoxically make Israeli nuclear weapons more rather than less likely; yet for the Middle East, possibilities of this sort must be considered. Hence it also can not be certain that Israel would receive any necessary assistance on desalination plants or nuclear facilities even after she had adhered to the NPT. At the least, the willingness of America and other countries to come across with this kind of aid would be one more item of discussion on any bargaining agenda.

ISRAELI BOMBS

Israel thus has had marginal reasons to reject the NPT; perhaps she should on similar considerations contemplate making nuclear weapons. Eventually, or even under present circumstances, the Israeli government will have to ask itself, How close is too close? At what point are peace and the national interest better served by actually going ahead to produce some bombs rather than always being three weeks from having an arsenal? A three-week time-lag may be a built-in inducement to someone else's preemptive attack where a hardened second-strike force would not. Staying years away from weapons, for example, by never building a plutonium separation plant, might thus be the most magnanimous arms-control policy. If commercial considerations or patriotic military prudence require the building of such a plant, however, military stability might be better served if Israel then went ahead to stockpile some bombs where Arab preemptive attacks could not get at them.

One could all along have conjured up battlefield scenarios in which Israeli nuclear weapons made a significant difference. The classic Israeli fear used to be of an Arab conventional armored attack that would sweep down on Tel Aviv and Haifa to push the Israelis into the sea. This also has often seemed to be the Arab dream. The Israeli expansion of 1967 may have made such an eventuality seem less likely, but one can still speculate about it in case forty million Arabs ever become as technically and militarily competent as two or three million Israelis. To remove the Israeli nightmare, or the Arab dream, one might add a small nuclear retaliatory force capable of destroying five or six Arab cities and the Aswan High Dam. Would this not force the Arabs once and for all to give up hopes of driving Israel into the sea? Wouldn't it mean, in Gallois fashion, that the Arabs could never exploit any battlefield victory they might ever win, for fear of the intolerable last gasp retaliation the Israeli nuclear force would inflict?

Financial costs per se are far from being a decisive obstacle; Israel's defense budget, despite the victory of 1967, has had to rise to in excess of one billion U.S. dollars per year.[3] By comparison, the annual costs of a modest plutonium bomb program, one capable of producing some ten Nagasaki-sized bombs per year, have been estimated in U.N. Secretary General Thant's 1968 report as about nineteen million U.S. dollars.[4] The current Dimona reactor has been producing enough plutonium for one bomb a year; six bombs produced and stockpiled over six years would not necessarily be an insignificant number for the Middle East.

---

[3] Institute for Strategic Studies, *The Military Balance: 1970-1971* (London, 1970), p. 40.

[4] *Report of the Secretary-General on the Effects of the Possible Use of Nuclear Weapons and the Security and Economic Implications for States of the Acquisition and Further Development of These Weapons* (New York: United Nations, 1968).

What of a guarantee by one of the existing nuclear powers, specifically the United States, of an alternative to Israeli atomic bombs? The United States might very plausibly retaliate for any nuclear attack on Israel, but such nuclear attacks might not be the most pressing threat. Does the American nuclear umbrella extend to deterring Arab conventional attacks on Israel? The answer has hardly been certain. Washington would very probably escalate to the nuclear level to defend West Germany against Russian armor, but very possibly might not escalate thus to defend Israel against the identical model of Russian tanks driven by Syrians or Egyptians. For the moment, the risks of Israel being defeated on the conventional war battlefield seem so low that the United States does not have to give much consideration to its position here; in any event, there is little evidence that Israel is seeking this kind of American guarantee or could expect it to be swift enough to be effective. Lack of security, not lack of guarantees, may have been a more honest Israeli argument against signing the treaty.

Israel's spectacular victory of 1967 could be interpreted in conflicting ways. At first Israelis saw it as a display, once again, of the superior military prowess of their soldiers and thus saw no need to change the rules of the game. After some days, details emerged on the Israeli preemptive air strike which established absolute air superiority from the very beginning of the ground campaign, suggesting that the victory was more heavily dependent on a full and fortuitous exploitation of technology, which might thus imply that Israeli victories in the future will depend on every technological avenue (even the nuclear avenue) being explored. More sober analyses however show that Israel would have won the 1967 war in any event, even had the initial air strikes not succeeded, but at a slower pace with higher casualties. For the moment, Arab victories in a tank battle are not so imminent.

For the Arabs to sweep the battlefield in the foreseeable future, massive Soviet assistance and participation would be required, which sets the stage for the nuclear-weapons question as Israel now must really see it. If the Soviet Union were really to offer all-out assistance to one or another of the Arab states, the Israeli decision would be easy, for nuclear weapons would be the only means of preventing or deterring military defeat. If, on the other hand, Soviet manpower and weapons assistance were not available to Syria and Egypt at all, there would most probably be no sense in Israel doing more than it did before 1967, that is, leaking occasional rumors about nuclear capabilities. The ins and outs of post-1967 Soviet assistance to the Arabs leave the situation somewhere in between.

At no point has it seemed likely that the Soviet Union would commit its troops and armored forces to a campaign of extermination against Israel. Any such intervention has at least been constrained by Russian aversions to baiting the United States into a direct confrontation. Rather the Russians have tended to offer the kinds of aircraft and weapons systems which reinforce Arab defenses without greatly augmenting their offensive capabilities. Even such levels of moderate assistance might still over time have driven Israel to seek nuclear weapons, if no other events had emerged to break the pattern. Any visible Soviet assistance works to keep alive Arab dreams of an eventual recovery of all of Palestine; and it might be that Israeli nuclear weapons alone can serve to terminate such dreams.

If the artillery barrages across the Suez Canal in 1970 had continued without letup, Israel might thus well have seen great advantages in going nuclear. Yet Soviet support for a warlike Arab posture has not been all that steady, and events such as the death of Nasser and the ouster of Russian forces by President Sadat suggest that the chain of endless Arab hostility might still be breakable. Only the most optimistic observer would see a Middle East settlement coming without the most prolonged and torturous diplomatic maneuvering, but those seeking a "new way out" of Israel's defense problems for a time can fix their attentions on how best to exploit Nasser's departure, rather than on nuclear physics. To detonate or otherwise publicize a bomb would now seem to be risking whatever chances might have existed for a peaceful settlement.

Yet the days may be gone forever when one could discuss the military balance in the Middle East without saying something about nuclear possibilities. Israel's recent technical progress in this direction has seen to that, as has the commitment of a nuclear power, the Soviet Union, to the Arab side. One must indeed ask why the Soviet Union has not committed itself as fully to the Arab cause as Cairo once may have hoped, or Israel feared. Was it just because of fears of U.S. counterintervention, or precisely because of the nuclear options Israel might have begun to exercise? The Israelis assuredly do not wish Soviet forces to play a more active role alongside Arab forces, and deterring this may thus at last have given a real purpose to any potential Israeli nuclear force. Yet such a force would have to be more formidable if it were the Soviet Union that had become involved on the other side. While delivery systems are no problem within the Middle East, they become more expensive if one wants to threaten targets in Russia. Five or six bombs might be a sufficient stockpile with which to confront Egypt, but not the Soviet Union.

Israelis have generally been reluctant to speculate about wars in which they have to fight Soviet forces. Such wars have been too far into the future and too unpleasant to contemplate; yet some speculation about such combat may now be unavoidable.

HOSTILE BOMBS

Israel will not bear all the blame for introducing nuclear possibilities between itself and the Soviet Union. It is certainly plausible that the greatly expanded Russian navy in the Mediterranean has some nuclear weapons on board, if only because its American counterpart has been widely advertised as possessing them. There is nothing in the NPT to preclude this, or even to preclude the deployment of Russian nuclear weapons ashore in Egypt, as long as they are not placed under non-Russian control. Here the Israelis are correct in their statements that the NPT settles nothing for the Middle East.

The presence of the two navies in the Mediterranean thus brings "nuclear umbrellas" (as well as "nuclear rain") more visibly into the picture. There have been no such navies as directly off the coasts of Japan, or India, or Australia. Access by sea may be a blessing if it means that tactical nuclear weapons can be kept around without deploying them on someone's territory, where they would be surrounded by his troops. But this access can also generate some drawbacks if it tempts the superpowers to intervene in ways which upset the stability of the region.

According to the text of the NPT, the Soviet Union stands committed not to give the Arab states nuclear weapons, no matter what Israel does (even if Israel makes bombs). According to the political liturgy which accompanied superpower efforts to sell the treaty, however, the Soviet Union stands morally obligated to provide extensive assistance to any non-nuclear state threatened by any nuclear-weapons state. An Israeli bomb could thus have served to deepen Soviet involvement on behalf of the Arabs, perhaps with Russian armored forces being deployed to Egypt and Syria. It is also possible that fears of Israel going nuclear have in part deterred the Russians from so explicit a deployment in the past. Undeployed Russian troops and unmanufactured Israeli bombs may thus deter each other.

The balance of trade-offs could instead rest at, or shift to, an entirely nuclear level. If Israel were to announce itself as the sixth member of the nuclear club, the Soviet Union might have to match this somehow; if the Soviet Union decided to brandish its own nuclear weapons in any fashion, either by rocket-rattling from the Soviet Union itself or by deploying visible nuclears forward to the Mediterranean or Syria, the Israeli public would take up a more serious discussion of its nuclear-weapons option.

Egypt apparently signed the NPT under Russian pressure and would presumably be persuaded to ratify if Israel were to do so. There have been press reports that the Russians rejected Egyptian requests for assistance on nuclear weapons, even requests for assistance contingent only on Israel having first gone nuclear.[5] It thus seems that Moscow still wants the U.A.R. and the entire Middle East to remain non-nuclear.

Yet this does not necessarily convince the Israelis that they are passing up an option to keep the Middle East free of nuclear weapons. Peking has not signed the NPT, and delegations of physicists from Egypt occasionally pay well-publicized visits to China. Earlier, China had made noises about the advantages of proliferation. Israelis in favor of retaining a nuclear option argue that there is no assurance that Egypt will not quietly receive such weapons from China, even if the Soviet Union can be trusted. If the U.A.R. were to gain control over such weapons when Israel had none, the weapons could easily come into use, either because the Arabs somehow expected thereby to win a war or because revenge for past defeats was now possible, even without victory, by bombing Tel Aviv and Haifa.

---

[5] *New York Times,* February 4, 1966, p. 1.

Israel could have been seen as immune to nuclear attack because the Arab population of Israel and its conquered territories constitutes a hostage that would suffer too much if any Arab states were to use such bombs. This way of looking at the situation of course assumes a greater rationality on the part of the Arab leadership than its superficial style would sometimes suggest and a very genuine Russian identification with the Arab Palestinian. Yet the Arab-Israeli conflict is at heart a conflict over real estate; Arab refugees never will admit to less than an intention to go back to the lands where they were born, to resume life there as before. If nuclear weapons were used against Haifa, Acre would suffer also. Could any Arab use nuclear weapons against Jerusalem knowing what this would do to the old city, its people, and its holy places (Moslem as well as Christian and Jewish)? If Tel Aviv offers a more purely Jewish target for attack, still it is the area to which the former residents of Jaffa wish to return. Israel is a small country, and the winds that spread nuclear fall-out would hardly spare the refugee camps on either side of the Jordan River. Aside from the Israeli nuclear complex in Dimona and some concentration of armor in the Sinai, therefore, the list of targets Cairo or Damascus would want to hit might seem short.

Since we all have known about the atomic bomb since 1945, it is not easy to conclude that Egypt would use it against Israel at the first opportunity or even that Cairo so seriously contemplates the "eradication" of Israel. If no other evidence came to mind, one would at least have to consider some implications of the construction of the Aswan High Dam. The dam clearly promises some great benefits for the U.A.R. But the dam also couples a great deal of destruction down the Nile Valley to any single atomic bomb which could breach its walls; it serves to oppose any effort to make the U.A.R.'s population less vulnerable to nuclear attack. If Israel can thus plausibly come into the possession of only a few bombs within the next few years, has construction of the Aswan High Dam not signaled Nasser's clear recognition that he could never complete an eradication of Israel, even if his tank columns for once won the great land battles? If Egypt were somehow to come into the possession of nuclear weapons, does the dam at Aswan plus Cairo's compactness not offer the Israelis a sure target by which to deter any Egyptian use of such nuclear warheads?

As suggested earlier, however, this may all impute greater rationality to the Arab leadership than exists. In a broad sense, it could indeed be argued that the persistent Arab quest for justice vis-à-vis Israel has been suicidal; a wild use of nuclear weapons might at least seem to offer retribution and revenge if the Arab states still felt frustrated and cheated, even if the Israeli retaliation would be just as wild and severe.

AMERICAN RESISTANCE

If the immediate requirement for nuclear weapons is not yet overwhelming, there are still serious arguments against Israel acquiring them. The alienation of the United States is clearly an important consideration, perhaps the most important. It would be one thing for Israel to refuse to sign the NPT, another to go ahead and violate it. Such a move would shock and perhaps even immobilize the American Jewish community and would generally antagonize much other pro-Israel opinion in America, as well as in Great Britain and Europe. No one can definitively predict the U.S. governmental reaction. If an American administration were at all anxious to disengage from commitments to Israel, such an action might be justified in the eyes of a world that is resentful of Israeli nuclear-weapons production. Conversely, if it chose to remain responsible for peace in the Middle East, the United States might nonetheless decide that such nuclear activity required severe retaliation, such as a freezing of American private monetary transfers to Israel, perhaps until Israel cancelled all nuclear-weapons programs and submitted completely to IAEA inspection, which would guarantee no resumption. If Israel were forced to submit in such a case, it would be worse off than when it began.

The logical place to draw the line against an Israeli bomb might have seemed to lie at the plutonium separation plant. Soon any advanced country may be able to make a commercial case for a plutonium plant, and thus the last plausible distinction between military and civilian production facilities will have been clouded. For the moment, however, Israel's commercial claim for such a facility can still be challenged. Israel, like India, could claim that this plant is already commercially necessary, if only to prepare fuel for future fast breeder reactors; but the suspicions of a move to weapons are strong in both cases.

Indeed American pressure has been quietly applied against any Israeli construction of a plutonium plant, during both the Johnson and the Nixon administrations. The United States has demanded and received the right to "visit" Dimona twice a year, although these "visits" are not to be officially labeled as inspections. Threats have not been explicit or public, but it has at least been hinted that the United States would treat this Israeli approach to the bomb as cause for some of the reprisals suggested above.

Yet all this pressure may indeed have failed, as the U.S. bluff is being called. Such failure illustrates one of the more serious problems in the world's general resistance to proliferation, once such proliferation assumes a de facto or "clandestine" form. The world is ready to back up a condemnation of proliferation, but it expects any indictment to be justified by evidence of distinctly "noncivilian" activity long before bombs come off the assembly line; when it consults its scientists, however, they advise that no such clear distinction remains on which world opinion can be based. (The world will also expect a very noticeable test explosion before weapons can be said to exist; here again it may be disappointed in ways which confuse the resistance to proliferation.)

If the last clear barrier to nuclear weapons was the lack of a plutonium reprocessing plant, it is possible that this barrier no longer remains. It is rumored that the plant has existed for several years now, in Beersheba rather than Dimona. Given the regular surveillance of the Middle East by American (and Russian) reconnaissance satellites and by U-2 overflights, it is likely that the United States would be fully aware of this plant, which enables Israel to prepare plutonium for use in bombs.

The United States may thus have been outflanked by the Israelis on the sanctions implicit for such a plant. As long as Israel does not publicize the plant's existence, the United States does not have to take a stand, unless it elects to publicize the plant itself. The tennis ball, in effect, is in the American court. If the plant's existence is announced, the United States will be faced with an embarrassing choice, given the continued availability of Soviet military assistance to the Arab states. Most of the world does not understand so clearly what role a plutonium reprocessing plant plays, or even what plutonium is. If Israel were to detonate a device under the Sinai desert, American public opinion and world opinion might indeed be transformed so that the Nixon administration could credibly withdraw from its Middle East involvement or alternatively impose financial or other sanctions on Israel. However, if the evidence is simply U-2 photos of a plutonium plant, Zionist opinion in the United States would hardly be neutralized and the world would wonder whether the United States, in the face of Soviet pressure, had not simply seized an excuse to back out of the Middle East.

THE SEMANTIC PROBLEM

It has been argued that Israel's indigenous sources of uranium have not been sufficient to produce more than perhaps one bomb a year. With a five-year backlog from the operation of Dimona, five or six bombs might thus now be available. Yet there are persistent rumors of Israeli access to much better sources of uranium, specifically in the Republic of South Africa.

If South Africa would not benefit from openly cooperating with Israel or from reaching for such weapons itself, this does not preclude immediate cooperation with Israel on such weapons. If the exchange were one of uranium for expertise, Pretoria could at least stockpile the expertise now, even if not to use it until the early 1980s. The uranium delivered to Israel need not be openly identified as such; and Pretoria's complicity can always be denied. The crucial question for South African respectability may hinge on whether a sixth country has detonated a bomb rather than manufactured one. Israel could assuredly produce enough plutonium, as it is, to detonate one bomb for test or demonstration purposes; however, it may not make good sense to do so. Pretoria could more easily encourage Israel to postpone detonation if it held the leverage of ongoing uranium sales.

If uranium is available and the plutonium reprocessing facility is already in operation, whether or not Israel "has the bomb" has indeed become a semantic question. Israeli statements do not necessarily pledge that no bombs are in Israel's arsenal. Premier Eshkol stated that his nation would not be the first to introduce nuclear weapons into the Middle East.[6] The pledge leaves ambiguities which hardly renounce all Israeli bomb programs, since nuclear weapons have already been introduced into the Middle East, on board ships of the American Sixth Fleet and very possibly on board Soviet ships in the Mediterranean. Also, bomb components can now be produced and accumulated in quantity without necessarily finally assembling them. Even American nuclear weapons are flown in configurations with enough of the rim missing so that they cannot explode accidentally. Is something that cannot explode "a bomb"? Perhaps American (as well as Israeli) assemblies can be called bombs only once they are in flight, when a crewman inserts the last component.

Politically the question will now be whether Israel can exploit nuclear weapons it cannot formally publicize. Is there any application of nuclear weapons in today's world which does not require advance warning? Most of the scenarios for Israeli exploitation of nuclear weapons are indeed deterrent situations. Perhaps one could envision an actual tactical use of such weapons against a large Arab concentration of armor in the Sinai desert, or some Russian amphibious fleet about to storm Tel Aviv in 1978, but the real value of nuclear weapons is in their use in intimidating or deterring Arabs and perhaps in deterring Russians.

To exploit nuclear weapons, therefore, Israel needs to publicize them, but only a little bit. If Arab newspapers persist in accusing Israel of manufacturing bombs, so much the better. If press reports in other nations make the same accusation, Israel presumably will not mind. Indeed there are indications that Israeli officials have encouraged such commentary, as in the case of a *Der Spiegel* article in 1969.[7] Such "leaks" serve at least two functions. They publicize the bomb somewhat, but they also test the world's reaction to see whether severe sanctions would flow from any brazen Israeli display of a nuclear-weapons arsenal. Did radical students demonstrate in front of the Israeli embassy in Washington after the *Der Spiegel* report? Indeed they did not. One plausible policy for Israel to pursue is simply to edge into a public acknowledgment of having gone nuclear, comparable to the Russian statements on the bomb in 1947 and 1948.

[6] Statement of Prime Minister Eshkol, as quoted in *New York Times,* May 19, 1966, p. 14.
[7] See the *Der Spiegel* article of May 1969, cited in *New York Times,* May 8, 1969, p. 15.

Deterrent signals do not have to be transmitted so openly either. Washington and Moscow can be informed privately of anything their reconnaissance systems have not told them already. It is rumored that in the spring of 1970 Russian officials told their Rumanian counterparts that Israeli had threatened nuclear war; as alleged proof of Israeli madness and unreasonableness, Israel is supposed to have threatened a nuclear attack on Kharkov and Kiev in the extreme event of an Israeli defeat. If one took this threat at all seriously, concern would shift to the delivery systems required to reach across Turkey to Russia from bases in Israel and to penetrate Soviet air defenses.

American policy on the supply of Phantom F-4 jet fighter-bombers to Israel has been extensively discussed in the context of tactical nonatomic operations in the Middle East itself. Yet the Phantom F-4 has sufficient carrying capacity for atomic weapons and an extreme range of 2,000 miles, enough to reach the southern Soviet Union. Allowing for the loss of range entailed in evasive maneuvers to by-pass air defense, the remaining range would still possibly be sufficient to penetrate the Ukraine; this introduces at least an extra consideration to the discussion of supplying airplanes to Israel.

It is reported that Israeli officials negotiating the purchases of the Phantom F-4 asked their Pentagon counterparts whether the airplanes might be equipped with racks suitable for carrying nuclear weapons.[8] This request was refused. It is difficult to believe that a country capable of producing nuclear weapons could not itself engineer the bomb racks required to attach them to an F-4; the request was thus probably intended to test American reactions to the nuclear question rather than to acquire any particular piece of military hardware. The United States government did not indignantly terminate the negotiations on the sale of aircraft because Israel had raised the nuclear possibility; Israel once again had its answer.

If one wanted to make a definitively clear announcement of one's bomb stockpile in this world, an actual detonation might still be required; and if Israel felt it had to conduct a test detonation for any other reason, it would of course be saddled with the costs as well as the advantages of such a definitive

[8] *New York Times,* July 18, 1970, pp. 1, 8.

announcement. Perhaps there are still generals who would refuse to accept an untested nuclear weapon. Israeli military officers, perhaps because of their lack of old traditions, have however shown themselves to be generally more sophisticated than the average in other countries' military hierarchies. It surely would be possible for an Israeli physicist to convince an air force general that the bomb was much more likely to explode than the "untested" bomb the United States dropped on Hiroshima. Computer simulations can test an explosive device (and rumors have it that Israel has indeed done such a simulation); whether or not this form of "testing" is fully equivalent to a real detonation in political terms is a different question.

The slightly less sophisticated public still believes that one must go out of his way to make nuclear weapons and that one must explode a bomb in order to have one. Any definitive condemnation of proliferation will thus be held in abeyance until a detonation occurs. No matter how conclusively our scientists tell us that the bombs must exist, detonations will be required to convince the public that there should be real sanctions against Israel.

Israel would be unique among possessors of nuclear weapons in that no other club member's homeland is immediately under threat. To some extent this reduces the need for brandishing such weapons openly. Other states brandish nuclear weapons, first, to prove they have them and, second, to make more credible the possibility of their use in marginal situations. There is less need for Israel to play games of nuclear brinkmanship; such weapons would indeed never be used for anything less than totally cataclysmic situations, specifically any use of nuclear weapons against Israel or any serious military penetration of Israel by Arab or Russian armored columns. No Israeli government would ever risk trying to couple threats of nuclear retaliation to Al Fatah guerrilla attacks or to Egyptian artillery barrages across the Suez Canal. If nothing else, this would give the maddest leader of the smallest Palestinian organization undreamed-of leverage and power. There was a time when U.S. nuclear testing was thought politically advisable in general terms to maintain the credibility of U.S. nuclear escalation in defense of West Germany. What satellite or marginal area exists for Israel such that "test" detonations would be required to establish an escalation credibility?

Nuclear weapons can serve two specific functions for Israel, neither of which really requires as much demonstration and explication as in the case of the five nuclear powers to date. The very potential for such weapons can inhibit the Arab states which lack such scientific prowess. The nuclear threat *in extremis* can force Arabs or Russians to forgo hopes of "wiping out" Israel.

It is thus entirely possible that Israel will become the first example of quasi-proliferation, understood to have nuclear weapons ready to fire but never having proved this to those in the world who require a detonation as proof. It is clear that a detonation makes some considerable difference, as it has in the past. Other members did not really join the club until they had exploded one of their bombs. One can indeed find fully qualified nuclear physicists from country to country who calmly state their belief that Israel already has bombs. Their calm demeanor might be taken as evidence that they have not really internalized their professed belief and are merely making statements for their conversational value; alternatively, the calmness could reflect the absence of an outside-world reaction. The outside world may never get excited about Israel having become number six until a bomb is detonated. Israel may conclude that all it needs to gain can be accomplished without such a detonation. If the intention were only to intimidate Arabs, rumors can be enough, perhaps even more effective than confirmed reality. If the intention is to deter Russians, signals can be transmitted clearly enough that another *force de frappe* exists *in extremis*. If the intention is to avoid testing American tolerance, then the Americans and the world should be given nothing so tangible as an explosion.

6

Tokyo

In February 1970, Japan signed the Nuclear Non-Proliferation Treaty, but the world was hardly reassured. Signature in 1968 would have suggested a ratification soon to follow. The signature in 1970 was instead accompanied by statements that ratification would not come quickly and that indeed no decision was implied by mere signature of the treaty.[1] Several political developments simply made signature tactically appropriate at this time: West Germany's signature of the NPT in December 1969 might otherwise have focused too much embarrassing attention on Tokyo's hesitation. The United States agreement to return Okinawa suggested at least some reciprocal gesture to substantiate the government's stress on close Japanese-American cooperation. Finally, negotiations on the exact form IAEA safeguards would assume for countries subjected to the NPT were to begin soon, and, as a state which had at least signed the treaty, Japan would probably be able to more effectively exercise leverage in these negotiations.

Signature was thus advisable, and it was all the easier because of several factors decoupling the normally speedy ratification process. Since July 1968, all Euratom signatures have been accompanied by provisos stating that ratification would not follow until specific agreement had been reached on the accommodation of Euratom and IAEA inspection procedures. Most of the Arab

[1] *New York Times,* February 4, 1970.

states signed the NPT immediately after it was offered but will withhold their ratifications until Israel accepts the treaty, which may be never. Even the two superpowers delayed their ratifications for more than a year, suggesting again that one can reasonably sign a treaty without being committed to its ultimate ratification. Japan does not have a history of considering ratification separately from treaty signature, but its eighteen-month delay before signing and instances of other countries delaying ratification suggest that Tokyo could now perfectly well reject a pact already signed. With an American president less enthusiastically in favor of the treaty than his predecessor, with growing economic and nationalist sentiment opposed to the NPT, a ratification will depend on special and fortunate circumstances prevailing in Japan's international environment in the next few years. The NPT has been neither dead nor assuredly alive in Japan.

Almost no one in Japan is at all enthusiastic about the NPT. All opposition parties have taken stands criticizing the treaty. A significant part of the governing Liberal Democratic Party (LDP) is more quietly unhappy about the treaty. Public opinion, to the extent that the public is aware of the issue, is negative. So also is business; so also are Japan's major newspapers. Wherefrom springs so much reluctance and opposition when Japan, being either protected by the American nuclear umbrella or unprotectable by any means, has no imminent need for nuclear weapons? Given Japan's "nuclear allergy" in the aftermath of Hiroshima and Nagasaki, why is the NPT not made-to-order for Japan? The Japanese response is interesting, in part because it illustrates feelings that may show up in other nations as they reach advanced stages of economic development.[2]

As late as 1965, an NPT would have sailed through ratification in Japan with as little discussion and resistance as the Test-Ban Treaty of 1963. A step toward disarmament, a product of new Russian-American understanding, such a treaty might have been endorsed by most if not all the opposition parties in Japan. As late as 1969, opposition within Japan had really only shown itself among representatives of the Japanese electrical industry, urging more caution before Japan committed itself to an arms-control

[2] For a discussion of some of the same questions on the NPT from a Japanese point of view, see Ryukichi Imai, "The Non-Proliferation Treaty and Japan," *Bulletin of the Atomic Scientists* (May 1969).

program that might be economically injurious; yet the opposition of such industrialists has supplied the essential core around which other layers of resistance now have been able to form.

COMMERCIAL OBJECTIONS

Industrial objections to the NPT have tended to hinge on the IAEA safeguards required in Article III. If nothing else focused Japanese attention on the safeguards question, it was the delay in inserting an agreed Article III into the Soviet-American draft and the debate on Euratom versus IAEA procedures that occasioned the delay.

Japanese objections to external inspection take a number of forms.[3] To consider even a trivial aspect, there is the language problem: Japan conducts and records its scientific operations in Japanese rather than in English; thus it will take time and effort and will cost money to translate and maintain duplicate records for non-Japanese inspectors to review.

More broadly, there is an inherent conflict between greater assurance to the outside world that no bombs are being produced and economically efficient operation of power reactors. Hints of such a conflict have already arisen. Most physicists estimate that only about 2 percent of the plutonium produced in power reactors would inevitably be uncontrollable; but Japanese electrical power company officers have suggested that anything greater than 90 percent certainty in inspection will constitute an unbearable economic burden. To some extent one cannot watch something without changing it. At the extreme, to measure the exact output of plutonium in a reactor, or to be certain of how much uranium is in its core, one might have to shut the reactor down from time to time or otherwise complicate its operation. While it might have been clearly acceptable to shut down a research reactor in the early days of Japanese nuclear activity, first under U.S., British, or Canadian inspection, then under the IAEA, such a shutdown would be intolerable if it interfered with Tokyo's electrical power production.

---

[3] A statement of the grievances of Japanese industrialists on inspection by the IAEA can be found in *Atoms in Japan* (Japan Atomic Industrial Forum, Inc., 1968), pp. 3-5.

Now that IAEA inspectors have their first large power reactors to monitor, anticipations of dissatisfaction have emerged among Japanese industrialists. An inspector who is congenial and spends half the morning sipping tea is one thing; one who insists on seeing all records or all fuel stocks is another. Fears of commercial espionage are also sometimes cited, because of the likely temptations to sell information on commercial-reactor designs to rival companies outside of Japan. For the moment, however, this is a somewhat abstract possibility, since some years will have to elapse before Japan begins designing and producing any commercially competitive reactors for export.

Reactor shutdowns and commercial espionage may be rather symbolic and unlikely bogies, but other aspects of inspection still promise to reduce monetary profits now that atomic energy has become a commercial enterprise. The costs of inspection may always be trivial compared to the defense budgets of the various states; yet in the accounting of a particular firm such costs may constitute a drainage of profits few corporation presidents could justify to their stockholders.

The mere uncertainty of emerging industrial prospects makes it very difficult to predict how costly monitoring and supervision will be.[4] For the future, Japanese industrialists envision themselves constructing enrichment plants for the reprocessing of plutonium and then fast breeder reactors. If today's power reactors can indeed be adequately policed without seriously handicapping the production of electrical power, it may be much more difficult to achieve a 90 or 98 percent degree of control on fissionable materials in the future. As the quantities of fissionable materials handled increase, moreover, even a 2 percent leakage may suffice for a politically significant clandestine weapons stockpile.

To refer to a relatively secondary consideration, although IAEA procedures allow a nation to reject any particular individual as an inspector if it feels he is prone to commercial espionage or obstructive legalism, Japanese officials have never seemed to find this provision particularly reassuring. The explanation is interesting. Throughout its relatively short diplomatic history, Japan has apparently been extremely reluctant to expel diplomats as *persona*

---

[4] A good basic discussion of the possible problems on safeguards is presented in Arnold Kramish, "The Watched and the Unwatched," *Adelphi Papers*, vol. 36 (June 1967).

non grata. Whether or not this reluctance stems from a special concern for the dignity and possible humiliation of the individual, it explains some of the concern about inspection procedures.

No one can argue against all inspection. No government is going to allow private electrical power companies to dispose of fissionable materials as they wish. Thus Japan must inspect and monitor its atomic energy industry, just as the AEC monitors that of the United States. Yet there are some significant differences between national inspectorates and those which would most probably emerge from the NPT. The national atomic energy agency is not only a mode of control but also a potential source of technology and support. Inspection by those of the same nationality, by those who also do some of the work, is much more acceptable, psychologically and ideologically, than inspection by outsiders who only ask questions. When safeguards on British-supplied equipment were handled on a bilateral basis, the same British personnel served as both inspectors and technical advisers, a much more popular arrangement in Japan than having the IAEA as the safeguarding body.

A number of Japanese have thus been urging a much more permissive standard for all internationally operated safeguards, a standard sufficient only perhaps to force any government wanting to make a bomb to do so deliberately and fairly explicitly and to have a responsible and effective national accounting system. Countries having friendly relations with their neighbors (such as Japan) might not have to undergo probing inspections of their nuclear facilities if clandestine manufacture of weapons were highly unlikely. Yet it must be acknowledged that countries having confrontations such as those between the Arab states and Israel might require much more than an effortless 90 percent degree of certainty, and the establishment of an IAEA double standard to distinguish between conditions in Israel and conditions in Japan will be politically difficult.

Japanese grumbling about inspection is a little misleading in any event. Unlike the Euratom countries, who had been inspecting themselves and could have hoped to continue to escape outside monitors, the Japanese would have had to expect inspection in any event for the immediate future, at least until the emergence of Japanese-produced equipment utilizing Japanese-produced fuel. With Japan's lack of uranium deposits, the latter was unlikely, except on a very small scale. Japan has not been accustomed to any self-inspection system, compared to which IAEA procedures might constitute a new and unpleasant nuisance (all reactors currently in Japan have been safeguarded either by the donor nation or more recently by the IAEA).

To some extent, then, the Japanese have been echoing German and Italian objections to the NPT, objections meaningful for Japan but not relevant to the NPT. By objecting to what the NPT would impose on Germany, the Japanese are in effect complaining about what would have been maintained in Japan, NPT or no. For the longer term, the NPT would have committed Japan to accepting a safeguards system even where it might otherwise have escaped it, which possibly accounts for Japan's suggestions for a time limit to the NPT.[5] At about the end of the five-year limit that Japan was proposing, the first Japanese facilities free of inspection under the old rules might have been just coming into operation. Suggestion of a time limit can be interpreted along other lines, for example, it might be Japan's contribution to the anti-NPT coalition's battery of arguments against the treaty, in which each recalcitrant nation stressed an objection which the others could then echo and cite. In truth, the positions of the recalcitrants are dissimilar enough to make it difficult to stress all the arguments jointly and evenly.

Japanese endorsements of Euratom objections to IAEA procedures indeed are only one side of the bargaining coin, for Tokyo shares interests with both camps of the Brussels-Vienna argument. Tokyo desires Euratom to win as many exemptions as possible, as long as Japan gets to share in each and every one of them; but it is determined to oppose any special privileges for Euratom that other advanced states would not share.

The counterargument of Europeans, and even Americans, is that Euratom involves control by adversaries which could not be produced by a purely national Japanese inspection system, no matter how technically proficient the system. Belgians can be counted upon to distrust West Germans, but can IAEA or the

---

[5] For the official statement of the Japanese position on the NPT and on imposing a time-limit on its duration, see United States Arms Control and Disarmament Agency, *Documents on Disarmament, 1968* (Washington, D.C.: U.S. Government Printing Office, 1969), pp. 309-14.

world really count on Japanese to monitor and distrust other Japanese? Some very vague speculation has emerged on the possibility of Australia and Japan, plus some other states, forming a consortium to control the uses of peaceful nuclear activities in some manner sufficiently "joint" to win whatever special treatment the IAEA accords Euratom. Yet the difficulties of getting such a multinational arrangement going are not trivial. It probably could not be arranged in time for the legal initiation of the NPT. For the moment Japan must be counted in the opposition to any treatment of Euratom which differs from that applied to national inspection systems; if Euratom does win special concessions, Tokyo might well find this a cause for delaying or rejecting ratification of the NPT.

In the earliest days of the development of the IAEA safeguards system, the Japanese government generally favored such endeavors, encouraging the IAEA to be aggressive in accepting responsibilities from various bilateral arrangements. It might thus seem ironic that the Japanese electrical industry is now reluctant to see such inspection extended or made permanent; but perhaps it demonstrates that the government did not consult its industrial sector extensively enough to protect its national interest. Some industrialists are in fact prepared to accuse the Foreign Ministry of not having raised objections early enough. Yet again Japan's relative slowness in developing its nuclear industry would have made a determinedly resisting Japanese posture somewhat premature in 1962 or even in 1967 or 1968. Rather more logical in terms of Japanese self-interest would be temporary acquiescence in inspection and controls followed by demands for complete review at about the time an indigenous Japanese program became feasible. Hence the suggestions for a five-year time limit on the treaty or for periodic review of its workings.

Thus the economic arguments against signing or ratifying an NPT have been the vanguard for Japanese opposition. The arguments have had a precisely material content or have reflected a more general feeling that Japan's political or property rights were gratuitously being signed away. The latter style overlaps in peculiar ways with a broader range of political and military questions, drawing in some nationalistic elements previously submerged in the aftermath of World War II.

POLITICAL OBJECTIONS

The exposure of serious political differences between the United States and Japan in 1971 shocked many Americans who had come to see Tokyo as a relatively faithful and contented ally. The very assertion of an "independent" Japanese stance on political and economic questions almost seems to violate the antinationalist spirit inculcated in Japan during the American occupation that is enshrined in the postwar constitution. However, closer observers had considerably earlier noted signs of Japanese inclinations toward a more "normal" foreign policy, a foreign policy that was not automatically cooperative and agreeable, that could be selfish on questions of exports and imports, that would not simply follow the lead of others on issues such as the handling of Communist China. Such a policy of course might also seek normalcy by the reestablishment of Japanese armed forces in order to play a role in the outside world.

As early as the late 1960s, when the NPT was being debated and proposed, there was thus already a widespread feeling among Japanese elites that Japan was not sufficiently acknowledged as a nation, that her feelings on various tangible and intangible matters were being ignored. The tangible issues have included Okinawa, the Security Treaty and the American bases within Japan that it provides for, Soviet occupation of the Kuriles, and, partially as a symbol, the Nuclear Non-Proliferation Treaty. Less tangible issues touched on whether or not Japan is automatically included in serious international discussions such as at the Eighteen Nation Disarmament Conference (ENDC), or even whether permanent Japanese membership in the U.N. Security Council is much discussed abroad.

By withholding signature in the summer of 1968, Japan clearly signaled that political concessions might be necessary to win Japanese approval of the NPT. The admission of Japan (together with Mongolia) to the Geneva disarmament negotiations in the spring of 1969 was a response to Tokyo's reluctance to become party to agreements it had no role in drafting. Yet by 1969 probably more than this was necessary to make the treaty acceptable for the Japanese as a whole.

Overall, Japanese political sentiment, left and right, was becoming increasingly indifferent to satisfying foreign standards of behavior. The stream now includes an ever more idealistic left pacifism which condemns the United States and the Soviet Union almost evenhandedly, an isolationism that urges Japan to consider primarily her own welfare, and a renewed nationalism or militarism which urges expansion of the armed forces and seeks overseas activities for the Self-Defense Force. As in other countries, a retained right to make nuclear weapons might seem to count for something even if it were never exercised. Even if no serious thought is ever given to the possible use of such weapons, the NPT seems to imply that Japan is taken for granted as acquiescent and not a great power, or as a permanently second-class power, the loser of World War II. As in other countries, abstract scenario-painting of the Gallois variety can conjure up wars in which the United States could not be counted upon to defend Japan, where Japan presumably might have to expend the greatest efforts to defend herself. Such speculation does not have to be serious to be significant in rounding out a nationalist line of reasoning opposing the NPT. Even the left opposition can echo some of these arguments. If one denounces the Security Treaty, one denounces with it any assumption that such a treaty would really motivate the United States to come to Japan's aid in a World War III. Most of those sounding these lines of reasoning will for the moment be content to retain a legal option of weapons manufacture; a few, of course, on the same nationalistic impulses, will want to go further and exercise the option.

The Japanese "nuclear allergy" has perhaps been overrated or taken too much for granted since 1945. There is no doubt still a great deal of revulsion to nuclear weapons in the aftermath of Hiroshima and Nagasaki, but there also has been a conscious exploitation of the guilt feelings of the United States and other nations with regard to the first and only use of nuclear weapons. Given Japan's need to reacquire respectability after World War II, it would indeed have been surprising if the Japanese had not encouraged Westerners' feelings of guilt about Hiroshima and Nagasaki. Hence there is still only a minority of Japanese who would even be willing to admit a desire for nuclear weapons for Japan—25 percent by 1968, which is a surprisingly *large* number. In the same poll some 50 percent *predicted,* resignedly, that Japan would at some point acquire such weapons, which perhaps suggests more accurately the limits of the real aversion to nuclear weapons.[6] And then younger people, who are less consciously aware of the horror of Hiroshima or Nagasaki, or less guilty about Nanking and Pearl Harbor, will be less reluctant to consider nuclear weapons for Japan.

The United States has hardly done all it could to reinforce the "allergy," which increases the general Japanese confusion about the NPT. If the United States had concentrated exclusively on selling the NPT, it should have tried to discredit all military nuclear equipment. But the United States has been reluctant to let Japan's "nuclear allergy" make it difficult or impossible for American aircraft or naval vessels to pass through the area. Emergency access to bases in Japan will be helpful for the defense of Korea if nothing else, and the extent of U.S. respect for Japan as a nuclearfree zone has thus purposely been left somewhat obscure. The United States has committed itself to consulting the Japanese government before bringing any nuclear weapons into Japan; no such consultations have been announced. U.S. nuclear-powered naval vessels have called at Japanese ports from time to time. Some observers charge that these calls have been more frequent than necessary, with the intention of curing the Japanese of their "allergy." The Japanese Left has demonstrated at the arrival of any submarine propelled by a nuclear power plant or any aircraft carrier with planes that normally carry nuclear warheads (the commitment to consultation does not require that American aircraft carriers divest themselves of nuclear warheads before calling at Japanese ports).

If the issue of U.S. military force deployments no longer pertains so directly to Japan, it has certainly been significant for Okinawa, which reverted to Japanese sovereignty in 1972. Okinawa, supporting a vast complex of American logistics bases and an important air base, had in no way been covered by the implicit American pledge not to deploy nuclear weapons within Japan. For a time American negotiators seemed anxious to win an exemption for American bases in the Ryukyu Islands, even after they reverted to Japan. The nuclear question was dropped later,

---

[6] For some very similar figures on Japanese public opinion on nuclear matters, see Yasumasa Tanaka, "Japanese Attitudes toward Nuclear Arms," *Public Opinion Quarterly* (Spring 1970).

more because of specific Japanese sentiments than because of a possible conflict with the logical tone of the NPT. The bargaining then was focused instead on American rights of access for conventional forces on the islands. At no point was the return of Okinawa suggested as a counterconcession to an acceptance of the NPT by Japan. America simply has not given that much priority to the treaty.

As with other nations resisting the NPT, a certain indifference to the views of the outside world now seems widespread among Japanese elites, both right and left, and probably also among the voting public. It is not surprising that the Japanese Socialist Party decided to oppose the NPT, along with demanding the return of Okinawa and the termination of the Security Treaty. With a "holier than all of you" attitude, the JSP can denounce both the United States and the Soviet Union for not including full disarmament in the treaty, and it can hope to win votes by doing so, since the Liberal Democrats are burdened with the unpopular task of pushing the treaty through. Attempts by Foreign Minister Tadeo Miki to negotiate a multiparty agreement not to make the treaty a political issue were fruitless or even counterproductive, for the drift of public opinion had already suggested that the NPT is a heavy cross for the government to bear. The more moderate Democratic Socialists, far less opposed to the American alliance, have also criticized the treaty. So indeed has Komeito, the somewhat right-wing opposition party of the Sokka Gakkai religious sect; and more surprisingly perhaps, so has the Japanese Communist Party, despite the Soviet Union's coauthorship of the treaty.

For the Japanese Communist Party, opposition to the NPT could serve a number of purposes. At a time when the Soviet Union was confronted with greater Communist dissension in Eastern Europe and elsewhere, the NPT was perhaps an easy issue with which to "defy" the Soviet Union. For Moscow the material cost of the defiance may not be so great, for JCP opposition can not by itself prevent ratification of the NPT. It is not even clear that the Soviet Union could or would pressure the JCP to adopt a protreaty position. If Moscow finds the opportunity to embarrass the ruling Liberal Democratic Party attractive enough, it might allow the JCP to win some domestic support. It may be that the Soviet Union really is concerned about getting Japan to forswear nuclear weapons; but Japan is not the most pressing substantive threat for the moment, and bullying the JCP on the question will be less politically profitable than bullying the LDP. The Japanese government hence will be under a twofold attack from the Communists on the NPT issue, by the JCP, which denounces the treaty, and by the Soviet Union, which periodically calls on Japan to sign it.

Even within the ruling Liberal Democratic Party, support for the NPT is anything but obvious. A television personality running as an LDP glamour candidate for the Upper House in 1968 won a record number of votes while criticizing the NPT and declaring that Japan had a significant nuclear future.[7] The unanimous unwillingness of opposition parties to support the treaty obviously frightens many within the LDP, but it encourages those in the LDP who hope to use opposition party stands as their own excuse never to ratify the treaty. With other unpopular stands to take, specifically on the Mutual Security Treaty, the LDP could thus confront the United States with a supposedly exhausted bank account of popular toleration for its programs, with the implication that any NPT ratification must be postponed.

Fears of economic disadvantage have been a real cause of Japanese resistance to the NPT, as well as an excuse for those wishing to escape the treaty for more political reasons. If the evolution of IAEA safeguards practices thus should reassure Japanese businessmen that they will not be subjected to excessive inspection and will not be denied any privileges won by Euratom, the prospects of Japan ratifying the treaty before 1973 or 1974 will have been enhanced. Yet politics can still take over where economics leaves off. The reversion of Okinawa has not gone as smoothly as might have been hoped, and the good feeling that might have been extracted from this return of territory may not suffice to help get the NPT ratified. Bitterness about restrictions on textile exports will now additionally strain Japanese-American relations. President Nixon's failure to consult Prime Minister Sato before sending Kissinger to Peking humiliated and shocked the Japanese. However few Japanese political leaders might be serious about wanting to acquire nuclear weapons in the foreseeable

---

[7] For a personality profile of Mr. Ishihara, including his views on nuclear matters, see *New York Times,* February 25, 1970.

future, quite a number will now welcome the political and psychological sense of escaping the pledge renouncing such weapons.

There is at least one strong argument for Japanese adherence to the NPT which may yet motivate the most responsible and moderate elements in the LDP to try to get it through. As mentioned earlier, Japan still will require large inputs of foreign technical assistance in the nuclear field, which might not be forthcoming if the NPT were rejected. IAEA safeguards might still be negotiated on a project-by-project basis, even if the NPT is not ratified, allowing Japan to approach nuclear self-sufficiency without signing away her military option for good; but fuel-supply problems would remain perhaps indefinitely. Moreover, while American technical assistance would still be legally possible under IAEA supervision even if the United States were party to the NPT and Japan were not, it is not clear that such assistance would be granted.

GOING IT ALONE

Yet the United States will soon have lost a lot of its monopoly position in the nuclear field; the prospect of other advanced countries rejecting the NPT may yet erode the support of those Japanese who now see the treaty as a necessary evil. One can not predict which nations in the end will accept the NPT and which will reject it; nor can one foretell the political and emotional processes that may be aroused by defiance of the great powers. It is at least possible that states that have rejected the treaty will look each other over after the dust has settled and perhaps decide to cooperate commercially and otherwise outside the NPT-IAEA system. If such a list includes India and France (which is almost for certain), South Africa, Brazil, Israel, and Japan, it is not absolutely clear that atomic research, peaceful or otherwise, will have been crippled for this group.[8]

---

[8] A comprehensive picture of the inevitable spread of nuclear technology in this region can be found in C. F. Barnaby, ed., *Preventing the Spread of Nuclear Weapons* (London: Souvenir Press, 1969).

The new Japanese indifference to world opinion works in several directions. Western commentators on attitudes toward the NPT have often speculated that Japan should try to avert the spread of nuclear weapons to India and Pakistan or to the Middle East. The need to impose an effective NPT in these areas presumably would lead Japan to submit herself to parallel controls. Yet the realities seem to be that there is little concern among any part of the Japanese political spectrum about the Middle East or India. Where Americans might see "brother Asians taking an interest in each other's problems," most Japanese show almost a complete indifference to whether India will sign the NPT or to whether India produces bombs or not. Japan might be invidiously embarrassed if too many other nations joined the nuclear club, but the immediate military risks of nuclear warfare in the Middle East or South Asia play far less a role. Acquisition of nuclear weapons by North (or South) Korea would raise tensions. If an NPT with safeguards prevented such an acquisition, it would have some redeeming benefits; but such proliferation for the moment depends far more on policies adopted by Communist China or other nuclear-weapons states than on the unchecked nuclear industries of smaller countries.

It now seems clear that India will refuse to sign the NPT. A certain possibility exists that India will not even adhere to the terms of the NPT but will go ahead to begin producing nuclear weapons. A question therefore remains as to whether Japan can be counted upon to deny assistance to India in weaponsrelated technologies, particularly rocketry and computers. Indian interest in Japanese rockets as potential military vehicles has already been suggested at several points, and it is rumored that India has offered Japan a test range facility in India for its rocket program, to spare Japan the costs of despatching ships to monitor its testfirings in an over-water range. For military purposes, Japanese rockets are generally understood to lack accurate guidance systems, which might inhibit their value for striking at any but the most sprawling targets. Yet India might still approach other sources (perhaps France) if it needed guidance apparatus for Japanese rockets; Japan has already sold space-research rockets to Indonesia and Yugoslavia (and also—according to some rumors—to Israel).[9]

[9] See Inz. B. Ruzicka's claim, "Kappa 6H," in the Czech technological journal, *Letectvi & Kosmonautika* 26 (December 1965): 30.

Given the commercial temptations involved, there seems to be little certainty that the sixth nuclear power will face any embargo from Japan on the kinds of goods that might significantly help a military program. A number of Japanese have referred to the notorious Japanese nonparticipation in the embargo on Rhodesia as suitably analogous to what could happen in the Indian case.

EXERCISE OF THE NUCLEAR OPTION

If Japan were to reject the NPT, production of nuclear weapons might still be unlikely, for the military situations in which such arms could be useful are not obvious. The most acceptable weapon for by-passing the "nuclear allergy" has been the ABM, but this will remain impractical as long as likely enemy missile launch sites are all so close to the Japanese homeland. Yet as the "allergy" continues to erode, Japanese scientists will definitely become more and more equipped to produce bombs when requested, and imaginative Japanese strategists may yet develop scenarios in which nuclear weapons would be appropriate. The 1970 Japanese White Paper on defense relayed a very mixed signal on the nuclear-weapons question. While foreseeing no move to such weapons in terms of current planning, the White Paper took pains to note that such weapons in defensive forms would not be incompatible with the Japanese constitution.[10]

For the moment, the navy, or the Naval Self-Defense Force, is most easily introduced into actual military operations. Its ships have for years undertaken long cruises to make courtesy calls in Europe and North America when parliamentary opposition would have precluded any external maneuvers by army or air force. Perhaps at American urging, it has begun visits into Southeast Asia. If anti-submarine warfare is to be the principal raison d'être for such forces, the possibility of nuclear warheads for depth charges has already been put to discussion within Japan. Like the ABM, such nuclear warheads could not be directly aimed at any civilian target and in some sense could only be used "defensively." If this use for nuclear weapons can be conjured up, so perhaps can

[10] The text of the White Paper can be found in *Survival* XIII (January 1971), pp. 2-6.

others. Like similar justifications all around the globe, they must always be taken only half-seriously, for the sheer glamour of handling such weapons will have an appeal of its own for many military professionals.

Very few Japanese cite a lack of security guarantees as the immediate reason for resisting the NPT, although other states use such an argument. The Mutual Security Treaty with the United States is quite specific, and Japanese nationalism tends to regard it as almost too specific.

The Soviet Union is not carrying on nuclear blackmail against Japan, and China has promised never to use nuclear weapons against anyone unless they use them first. Japan's high degree of urbanization surely makes it vulnerable to nuclear attack, but this urbanization makes a lack of American response all the more unthinkable. One can not find many tactical situations in which Peking or anyone else could apply nuclear weapons against Japan. One either kills great numbers of Japanese by hitting a city or one does not use nuclear weapons against Japan at all. With the special sympathy of the world behind it since Hiroshima and Nagasaki, Japan might even be better able to stand up to hints of such attack, as world condemnation would turn all the more readily against any foreign leader who spoke of "rockets falling on Tokyo."

Japan's non-nuclear status has implicit handicaps nonetheless when compared with nuclear China. If Japan produces the bomb, the world, for good or ill, might immediately come to see Tokyo rather than Peking as the real major power of Asia. GNP can outweigh manpower in the world's eyes; the sad fact is that nuclear weaponry can seem to outweigh either. But this is more a prestige argument than a security argument, except insofar as prestige can be translated into power and security.

For the present, any Japanese move to nuclear weapons runs the risk of severely alienating the United States and the Soviet Union. Even Communist China, in a reversal of its normally light-hearted endorsements of nuclear proliferation, has stated that under no circumstances should Japan contemplate acquiring nuclear weapons.[11] How these great-power attitudes will evolve

---

[11] See "No Nuclear Arming by Japanese Militarism is Permissable," *Peking Review*, April 19, 1968.

over time is problematical. If India, for example, defies the NPT sponsors and suffers no serious retaliations, then Japan may be more likely to do so.

Normalization of Japanese relations with Communist China might hold attention away from the nuclear-weapons question until the mid 1970s. Some members of each of the political parties in Japan hope that such relations can generate political and economic benefits for Japan. Yet there are also many real possibilities of conflict with Peking. The reversion of Okinawa itself introduces territorial disputes about various small islands to the north of Taiwan situated in what may be oil-rich areas. If the confrontation with China should thus be competitive as well as cooperative, it could all too easily lead Toyko to seek the political and psychological reinforcement of a nuclear-weapons program.

Any move to nuclear weapons also risks a severe domestic reaction from the Japanese opposition parties; yet an attack on the NPT may leave the opposition with its momentum going in too many wrong directions to effectively agitate against nuclear weapons. If a member of the LDP were deliberately trying to set Japanese public opinion up for nuclear-weapons acquisition, he could hardly do better than to have the Socialists defeat the NPT in a bitter campaign which saw both the United States and the Soviet Union denounced.

Any real Japanese desire for nuclear weapons will obviously also depend on an eroding American image in the Far East in the aftermath of Vietnam. If the United States makes only a moderate withdrawal from the region, less drastic Japanese responses will seem in order. But if the withdrawal from Vietnam leads to withdrawal from Laos, Cambodia, Thailand, Indonesia, and the Philippines, it will be difficult for Japan (or Australia) to accept assurances that America is as committed to their defense as ever.

Even if all of the underdeveloped Far East should fall to revolutionary movements, this would not pose any direct military threat to Japan. Yet the decline of American prestige would tempt many Japanese to fill an apparent military vacuum. Tokyo is not going to send ground forces to replace those of the United States. Stabilizing the politics of Southeast Asia through Japanese economic weight thus depends entirely on some other source (presumably the United States) checking the military ambitions of the revolutionary forces, to provide a screen behind which

Japanese investment can take place. In the event of a more total Communist military victory, the Japanese "replacement" of American power rather would take less committal forms—a bolstered naval presence which showed the flag extensively, but always a discrete twelve miles offshore, and perhaps the acquisition of nuclear weapons.

The combination of events is thus moving to make nuclear weapons more acceptable in Japan. Ratification of the NPT could head this off; if ratification were frustrated, however, probomb tendencies would be accelerated. Technology is working to make bombs easier and to allow a less explicit approach to military projects. If the government is openly reluctant to exercise the weapons-production option it retains after a rejection of the NPT, the question of clandestine or semiclandestine bomb manufacture might yet arise. The Japanese legal situation on atomic energy is a little peculiar in that firms engaged in nuclear work are required to state publicly the nature and purposes of their research. Some opponents of the NPT are prepared to argue that this is a sufficient guarantee against any clandestine military projects, but others have noted that firms are not required to disclose their "commercially valuable" secrets and that the distinction between military and commercial ventures is not so clear. A few commentators have feared that such openness as accompanies this law will allow other states seeking weapons to exploit even relatively innocent Japanese research. Given the likely reluctance of the Japanese Diet to change the status quo by ratifying the NPT, any move to change the antisecrecy laws also is not likely to be well-received.

The issue of the openness of Japanese nuclear activity is somewhat interestingly confused by the distribution of political attitudes among those Japanese with talents essential to bomb manufacture. As much as Japanese physicists in the university community might resent and oppose outside inspection, their domestic political outlook is still very leftist, decidedly anti-weapons, and quite distrustful of the intentions of the ruling LDP. It thus is not totally implausible that a Japanese clandestine weapons program would be exposed by Japanese physicists themselves and that self-inspection in the pluralistic Japan of the 1970s would be as reliable as self-inspection within Euratom. The attitude of the Left-leaning Japanese scientist thus affects the NPT in opposing ways; it concurs in the need for controls on the government but seems to make outside controls redundant.

Opinion polls indicate that aversions to nuclear weapons are declining among younger Japanese, presumably including younger physicists and engineers, who perhaps do not remember Hiroshima and Nagasaki; but those over age thirty-five remain firmly opposed to the bomb. The strongly hierarchical structure of Japanese university life holds a paradoxical advantage from a leftist point of view, as little tolerance has been shown any junior scientists expressing "open-minded" opinions on nuclear weapons for Japan. Whether the essential minimum of scientists required for a clandestine government bomb project would be small enough to escape this kind of intolerance may be a question that only time will settle. Increased employment of nuclear physicists in the electrical industry over time may weaken this academic control over junior physicists' probomb tendencies.

A Japanese decision to produce nuclear weapons is thus still a good distance off; even the ability to produce weapons is a little further off than a more alarmed view would have it. For the moment the more serious problem is the damage being done to any genuine domestic resistance to Japanese nuclear weapons. The process of debating the NPT has driven the opposition parties to create new ad hoc rationalizations of position where previously they would have come automatically. It has allowed some members of the governing party to profess to be in favor of renouncing weapons, while quietly contemplating the prospect of never having (or being allowed) to do so. When inevitable generational changes of attitude and inevitable technological progress are added to this, it is not really clear what would foster opposition to a Japanese weapons program in the late 1970s, except a firm adherence to the NPT now; yet this adherence can not be assured. Much will depend on who else signs and ratifies the treaty or on who defies it by actually detonating a nuclear bomb. If Germany ratifies but India detonates, Japan may watch the world's reactions closely before deciding how to act. If neither happens and the world stands still with many signatures and few ratifications, Japan probably will not be the first of these important states to commit itself.

7

Stockholm

Sweden may someday serve as the stellar example of the kind of psychological and political attitudes the Nuclear Non-Proliferation Treaty is meant to induce. A country which only twelve years ago was actively contemplating nuclear weapons today regards itself as a firm opponent of nuclear proliferation. Some changes of attitude here have been due to external developments which cast doubt on the military value of a Swedish nuclear-weapons program. Yet much of this is now also due to the international negotiations tail-wagging of the national interest dog, and the world might be very satisfied if this phenomenon could only be repeated in several more countries. Stockholm in effect has almost "been there" and has "come back."

THE TEMPTATION OF NUCLEAR WEAPONS

Swedish nuclear policy in the early 1950s required few "decisions," for "all that could be done" was already essentially underway as basic research. By 1957, however, certain choices had to be made, as additional funds might have accelerated a military capability and as research paths might for a time have diverged between civilian and military intentions. Earlier there had been a widespread assumption that Sweden might always be dependent on outside sources of nuclear technological support; the news now was that this was not really so. As a new option for Swedish defense could be discussed, the commander in chief of the armed forces had begun presenting programs for the ultimate development and production of tactical nuclear weapons. The attitudes of the electorate not yet being very clear, Swedish political parties were generally divided and embarrassed by the proposals.[1]

Nuclear weapons indeed seemed to make a great deal of sense for Sweden in 1957. Domestic physical and human resources (and reserves of uranium) could certainly support such a program, as they already supported the manufacture of sophisticated military aircraft; nuclear weapons were much closer to being "conventional" weapons then than they seem to be today. In the late 1950s, war was still a possibility that could be discussed seriously; armies might sweep back and forth across Europe, and Sweden might still have to force them all to respect her frontiers. Wars would still be fought with airplanes, much more than with missiles, and would take weeks and months rather than hours. The United States was widely touting tactical nuclear weapons as an equalizer which would favor the smaller battalions against the bigger, and there was little reason for Swedish military analysts to reach differing conclusions.

The threat of escalation to World War III, with attacks on the great cities of the Soviet Union, still seemed a crucial part of the American defense of West Berlin and West Germany. While no Swedish officers would publicly advocate that strategic reprisal weapons be brandished against Leningrad, the existence of Swedish nuclear weapons would nonetheless seem to increase the threat that aggression against Sweden could produce escalations so serious that Leningrad would in the end be destroyed; and any Soviet desire to reduce the legitimacy of nuclear weapons was undermined by Sputnik and Soviet ICBM tests, which certainly suggested nuclear warheads.

---

[1] A comprehensive account of Swedish nuclear decisionmaking in this period is to be found in Karl Birnbaum, "The Swedish Experience," in Alastair Buchan, ed., *A World of Nuclear Powers?* (Englewood Cliffs: Prentice-Hall, 1966), pp. 68-75. Aside from other sources cited, a special acknowledgment must be made to as yet unpublished seminar papers of Ingemar Dörfer.

Like other countries, Sweden would not have been indifferent to the prestige arising from the ability to manufacture nuclear weapons. A first-rate aircraft industry not only facilitates military defense within a framework of neutrality; it makes other people think more highly of Swedish economic and industrial prowess. Having the largest air force per capita in the world (indeed for many years one of the four or five largest absolutely) only bothers a few Swedes; to most—other things being equal—it has given pleasure. Other things remaining equal, similar considerations might have applied with regard to nuclear weapons.

At the same time, Sweden saw itself as a nation of peace, as one of the more humane and moral countries of the world; explicitly to procure nuclear weapons, which after Hiroshima and Nagasaki were not quite as legitimate as other weapons, might tarnish this image at home and abroad. If many might favor such weapons upon first discovering that Sweden could indeed manufacture them, others were opposed, while many remained undecided. A poll of Swedish public opinion at the beginning of the 1957 debate showed some 40 percent favoring explicit nuclear-weapons procurement, 36 percent opposed, and 24 percent undecided. The Conservative Party endorsed such procurement, while the governing Social Democrats, as well as the opposition Liberal and Agrarian parties, were divided on the issue; only the Communist Party was clearly opposed. At the close of the debate two years later, only 27 percent still favored acquisition of weapons, while 51 percent were opposed; yet many more than 27 percent, most probably a majority, still would favor retaining a Swedish option for weapons, which indeed was the policy the government sought to adopt.[2]

A SWEDISH COMPROMISE

The debate from 1957 to 1959 generated a compromise of sorts, which has faced little open challenge since, even when its premises have in fact become drastically outmoded. As in every

---

[2] Public opinion poll data can be found in Per Ahlmark, *Den Svenska Kärnvapendebatten* (Stockholm: Utrikespolitiska Institutet, 1965).

other near-nuclear country, two separate questions have been involved: (1) how to assemble fissionable materials into bombs and (2) how to procure the fissionable material in the first place.

Nineteen fifty-eight saw a postponement of any appropriations specifically for weapons development and a ruling that research be limited to "protection" against such weapons (in effect constricting such result as had been underway). By 1959, however, this ruling on research had been redefined again to tolerate a very broad notion of protection, even when it incidentally related to the techniques of bomb manufacture. In ensuing years such research clearly has answered many questions on the assembly of fissionable materials into explosives. Conservative Party efforts to force an acceleration and greater explication of this research program were defeated in a parliamentary vote in the spring of 1960. The implicit policy would remain implicit: to retain and accumulate a bomb option without openly exercising or accelerating it.

Programs purely for the manufacture of fissionable materials thus were not undertaken, but it was expected that a profitable nuclear electrical complex would emerge to generate flows of plutonium. If this complex were to utilize domestic uranium, Sweden would have a de facto bomb capability by the late 1960s, fully indigenous and free of external legal bars on military use. By the mid 1960s however, this last assumption had clearly been disproven, as foreign-supplied enriched uranium seemed a much cheaper route to electrical power, albeit under legal bans on any military use of the plutonium produced as a by-product.[3] As such facilities supplied the electrical power Sweden required, natural-uranium-fueled reactors would be redundant and could no longer be justified in civilian terms. Sweden's armed forces would not have access to bomb material without specifically and explicitly seeking it. In effect the bargains offered by the U.S. AEC had quietly undone a Swedish defense plan.

Swedish nuclear-weapons policy in the last ten years illustrates much of what is peculiar and fascinating about Swedish political style in general. Unlike most other near-nuclear states, Sweden does not often reward those who bring issues out into the open for

---

[3] A full discussion of these developments is to be found in Jan Prawitz, "A Nuclear Doctrine for Sweden," *Cooperation and Conflict* 13 (Autumn 1968): 184-93.

public discussion. The dominant theme of consensus and reason-ableness leaves less tolerance for "prophets in the wilderness," and the essence of political success is often to force one's opposition into a position where *they* would have to raise the public outcry to avoid surrender on the issue.

At times, new technological developments will force an issue into the open, as the public and the parties otherwise would find it difficult to tell where each stood. Such may have been the reasoning of those who broke open the nuclear-option question early in 1957. With the exception of this 1957-59 episode, which so seriously embarrassed all the parties, however, Swedish nuclear policy has been involved in competitive nonexplication, with the probomb factions the decided losers since 1961. Swedish public opinion today would severely punish any party that explicitly endorsed nuclear weapons, in large part because a respectable status quo has gradually developed which now seems to leave no role for such weapons. With the penalties for explication so large, one might sense a deceptive unanimity among the Swedish elite that weapons be forgotten. This might indeed be an exaggeration, for only a small percentage of those who came into the open twelve years ago have passed on or have been retired and only a portion of the remainder have completely changed their minds.

THE TEMPTATION OF DISARMAMENT

In effect, Swedish public opinion has turned against the military nuclear option. If an additional plutonium-production program would be necessary to fulfill the expectations of 1960, there has been no political agitation for such a program because the expectations of 1960 have been implicitly rejected. The explana-tion of this change in Sweden's idea of legitimate foreign policy must include both objective and subjective factors.

Many of the more clearly military arguments for nuclear weapons have indeed lost their relevance since 1960. With the acquisition of hundreds of missiles with megaton warheads by each side, the Swedish approach to "stay out of any war" had to change. General war would be a disaster for Sweden today no matter how much her neutrality were respected, if only because radioactive fallout has no respect for boundaries. In 1957 the threat might have been a Soviet or American move to turn a flank by a shortcut through Sweden. By 1965 such threats had lost much of their relevance.

The Kennedy administration entered office determined to "put the nuclear genie back into the bottle," and it did reduce the previous emphasis on tactical nuclear weapons. Yet it also expanded the U.S. missile inventory from about 50 to more than 1,600 and raised the deployment of American tactical nuclear warheads in Europe to some 7,000. It is difficult to say which of these changes had the most significance for Sweden, but the net impact was to weaken further any strategic military case for Swedish nuclear weapons. With 7,000 warheads perhaps deployed to Germany, would 50 Swedish tactical warheads really make a significant difference in great-power expectations for a World War III scenario? If an amphibious great-power force were about to storm the Swedish shores (a vulnerable target for any tactical nuclear weapon), was it really likely that such weapons would not come into use from the opposite great power's arsenal?

In the same period, work went ahead in both the United States and the Soviet Union on "defensive" antiaircraft systems, which inevitably make it difficult for ordinary bombers to penetrate to target. Since 1960, Swedes discussing nuclear weapons have had to consider delivery systems, an accessory that might still have seemed trivial in 1957. We often condemn the Soviet-American arms race of the last twenty-five years and sometimes scoff at the notion that this race might have discouraged other nations from buying arms; yet this clearly has happened in the Swedish case, as other nations' surface-to-air (SAM) missile systems have weakened the respectability of nuclear weapons within Sweden. The world was not that worried about Sweden; yet if the same phenomenon has been reproduced elsewhere, the great-power arms race will not have been a total waste.

Military arguments for proliferation aside, the world's view of the legitimacy of further proliferation has also changed, so that a nation which has not already explicitly committed itself to the bomb might find it more embarrassing and difficult to do so. Significant events here include a series of Irish resolutions at the U.N. General Assembly, calling for states not already having nuclear weapons to forgo them, resolutions which since 1960 have been supported by a majority of U.N. members. Also, the administrations of Kennedy and Johnson were far more explicitly opposed to the spread of nuclear weapons than the Eisenhower administration. Similarly, having accepted the legitimacy of nuclear weapons for themselves and the Americans in 1957, the Russians by 1963 had begun to deny it for any "nth" nuclear powers; the Test-Ban Treaty seemed to be a sign of this, and then, of course, so did the NPT.

Yet perhaps the most important influence on Swedish policy has reflected a procedural rather than a substantive development, namely, the addition in 1962 of eight neutral nations, including Sweden, to the disarmament negotiations at Geneva. Sweden's position as one of these eight at the ENDC has indeed had a significant impact on national considerations of military policy. The international role of Sweden in effect shifted from passive and circumspect neutrality to honest-broker arbitration, and public opinion adjusted to the change. First on the test ban and then on other disarmament steps, the Swedish delegation at Geneva saw itself as an independent source of expertise as well as spokesman for the other seven less economically developed nonaligned states. With this kind of vested interest in prodisarmament positions, a fuller national exploration of bomb options could well be forgone. If the Irish General Assembly resolution (and the antiproliferation implications of the test moratoriums and test ban) had not already reduced the legitimacy of Swedish nuclear weapons, Sweden's role at the ENDC had made them decisively obsolete in terms of prestige considerations. Rightly or wrongly, Swedes today see themselves as having worked for all forms of disarmament at Geneva, as having been against proliferation all along.

When the superpowers at last offered the NPT for signature in 1968, the nuclear-weapons option was hardly at the forefront of the Swedish imagination. If no other factors complicated the picture, Stockholm could indeed go ahead to adhere to a treaty renouncing this option.

POSSIBLE COMMERCIAL PROBLEMS

Yet to be against weapons per se is not the same as being opposed to all activities that might someday lead to weapons.

Some real commercial objections might have been raised to the surrender of Swedish nuclear-weapons options or to submitting to any foreign inspectorate to assure that weapons indeed were not being produced. Scientists in other countries are prepared to argue that military and civilian uses of atomic energy overlap so much that any constriction on one side would hamper the other. If nuclear energy is the industry of the future, the arbiter of who will be very rich and who will not, it might seem unwise for Sweden to accept a nonproliferation treaty that will handicap Britain and the United States so much less than itself.

Aside from constrictions on scientific curiosity, the NPT may require the kinds of inspection that take time and cost money, perhaps enough money at the margin to render certain electrical power projects unprofitable. If Sweden ever hopes to sell reactors on the world market, IAEA inspectors might be tempted to sell advance design secrets in commercial espionage, which General Electric can escape in the United States simply by labeling its test prototypes as "military."

Yet, despite the widespread objections to IAEA inspection in Germany and Japan, both having a nuclear status comparable to Sweden's, there have been very few such objections in Sweden. This divergence stems from a number of sources.

First, Sweden is a small country. Despite the fact that it has the second-highest per capita income in the world (or perhaps highest, if one counts certain kinds of social overhead absent in the United States), like other small countries it can not easily benefit from the kinds of economies of scale that probably will characterize the nuclear field. The role of large-scale exporter of advanced fast breeder reactors is probably less plausible for Sweden than for other countries, and thus the hypothetical threat of IAEA commercial espionage does not loom as large. Scandinavians in any event are abnormally honest, so that such international civil service malfeasance does not rest at the forefront of their imaginations.

If Sweden is to have a good share of the nuclear industrial boom, it would probably have to be as part of a larger market unit. If IAEA inspection is to replace Euratom under the terms of the treaty, that is an argument in favor of the NPT. While the procedures of Euratom are beneficial for the current member states, they generate trading obstacles for outsiders. Parallel but differing safeguards systems constitute a barrier to a free market for fuels and units of equipment; a single IAEA safeguards system would thus make it easier for Sweden to negotiate sales within the Common Market area. The Euratom agreement poses some other potentially obnoxious possibilities, whereby Swedish uranium might have to become available for French or European nuclear weapons. If Sweden is to negotiate for fuller commercial entry into Europe, it must either take a hard line on such questions or allow the NPT to supply the hard line seemingly from above. Clearly, the latter would make the Swedish task easier.

Swedish apprehensions about the application of NPT safeguards are further reduced by several less objective factors. The fact that the IAEA Director General is Swedish certainly helps, as Eklund is well-known to the management of Sweden's electrical and nuclear industries. The only Swedish experience with safeguards to date has been of a bilateral variety, with the United States and Great Britain; these inspections have gone very smoothly. Nothing in the Swedish experience so far confirms that an international bureaucracy is likely to be more officious and troublesome. Swedish industry moreover is less able to rouse public opinion, as compared, for example, with Japan. Whether government-owned or government-regulated, Swedish firms have long had to pay the social costs that accompanied their profits. If paper mills must forgo some returns to avoid polluting streams, nuclear electrical complexes may also have to pay some inspection costs to assure that they are not polluting the world with atomic bombs.

While the costs of inspections may be quite significant for the bookkeeping of an individual firm, they will be trivial by the standards of Swedish or other defense budgets. In principle, the Swedish government could reimburse the individual firm for any unusual costs in IAEA inspection; in fact there have been suggestions that this will be done.

There are technical experts in the Swedish nuclear establishment who indeed expect some serious costs in IAEA inspection procedures and some real difficulties in achieving high degrees of inspector-certainty that no plutonium is being diverted, that no bombs are being produced. Yet such experts are not given to sounding the alarm. The moderate-consensus tone of Swedish political life aside, such lack of alarm is reasonable, in that Sweden will not be the first country to be unjustly accused of possibly diverting plutonium. If the IAEA is going to decide that its mandate requires 98 percent certainty at some point, it will be in a country with suspicious and hostile neighbors—in Israel or West Germany.

SWEDEN AND THE TREATY

Given that the more normal substantive objections to the NPT are either outmoded or suppressed for Sweden and given Sweden's pride in disarmament activities and in its participation in the ENDC at Geneva, one might have expected the treaty to have clear sailing. Yet while the Swedish government seemingly had no national interests to block adherence to the treaty, it has sought to influence the treaty's development; Sweden still did not assign the NPT nearly as much priority as did the Americans or Russians. The treaty might displease knowledgeable Swedes today more for what it omitted than for what it included; if it were obvious that Swedish views and leadership had been frustrated or ignored, moreover, the final product would also have some serious drawbacks on procedural grounds, before a larger public.

For the government, or more particularly for the ruling Social Democratic Party and its ENDC delegation head, Mrs. Myrdal, the ideal would have been to bring home a treaty embodying some clear Soviet-American concessions that afterwards would have been recognizable as the "Swedish amendments." Possibly such amendments would have sweetened the treaty enough to get nations such as Japan or even India to sign more readily, and Sweden would internationally have been recognized as an effective spokesman for the uncommitted nuclear-industry states. If not—if no other states had really been impressed by the Swedish amendments—the final product would still have reflected well domestically Sweden's new influence in such serious matters. As it was, Russian and American coldness to such initiatives left the government with less than optimum choices: to admit defeat and reject the treaty or to bring home a treaty the Swedish origins of which are not overpoweringly clear. Given the Swedish people's impression that the NPT was a Swedish-designed treaty, the latter course probably suited the government's interest as the lesser evil, and the treaty was thus signed after a seven-week delay.

When asked why Sweden did not consider waiting longer before signing the NPT, Swedish officials generally respond that "[i]t would have seemed inconsistent for us not to sign when we had been in favor of the treaty for so long." This assumption of appearances is a little paradoxical.

It is significant that Swedes themselves see their government as having played a pro-NPT role and see outsiders as viewing Sweden in the same light. Yet among informed outsiders in 1967 and 1968, any list of NPT-recalcitrants would typically have included Sweden along with India, Japan, West Germany, Israel, and Italy. It is true that the Swedish delegation always endorsed the NPT "in principle," while suggesting some substantial modifications of the Soviet and American drafts. Yet India also favored the NPT "in principle," as did Japan and West Germany. If India's intention even to be bound by an NPT was suspect, so also therefore was Sweden's.

It clearly seems plausible that Swedish statements on the treaty make it easier for countries like India to abstain. Almost every other NPT-recalcitrant is somehow politically suspect—India because by 1959 it had already built a plutonium separation plant, Israel because of its immediate military problems and secrecy at Dimona, Germany, Italy, and Japan because they were the defeated Axis powers, Brazil because it was governed by a military junta. To have Swedish arguments to quote, to have Mrs. Myrdal's objections to cite, thus enhanced the respectability of otherwise suspect German or Indian statements. Regardless of whether the Swedish ENDC delegation sensed such developments, Americans and Russians clearly did, which accounts in part for their relative disinterest in Swedish proposals.

If Swedish efforts seemingly helped certain nations prepare their rejections of the NPT, a number of theories might emerge as to why Sweden did not instead match Canada in a more general backing of the U.S.-Soviet proposal. Remembering the earlier probomb agitations of 1957, some Americans might suspect that important elements in the Swedish military or civil service still wanted to hold open the military option. Others might charge the Swedish position off to a naive idealism which saw general disarmament to be just as feasible as a halt to proliferation.

The Swedish government may simply have felt more astutely

that it had some leverage here on the NPT question which could be applied to coerce the superpowers also into some disarmament. If Stockholm desired ratification of the NPT, Washington and Moscow seemed to want it more; elementary bargaining theory suggested that Sweden should not gratuitously surrender its asset. Since 1965, therefore, the Swedish ENDC delegation has consistently sought to couple any nonproliferation agreement to other disarmament steps, such as the comprehensive test ban or more significant great-power weapons reductions. Related to this, the delegation may even more astutely have sensed a minimum of concessions that other near-nuclear states would require before being reconciled to the NPT. By becoming the spokesman for the NPT-recalcitrants, perhaps Sweden could in the end lead them happily back into the fold.

If Swedes took this nonaligned leadership role at all seriously, however, it might not have been so compatible with the material or psychological interests of those being led. Several countries only welcomed the Swedish intervention as long as it indeed prepared their cases for abstaining from the NPT. If their erstwhile advocate instead seriously intended to negotiate a compromise answering objections and involving entry into the NPT system, this did not suit India or West Germany. The Swedish ENDC delegation conversely would be perturbed by any efforts by such powers to convert Sweden into simply another member of the anti-NPT coalition.

India, probably the most determined of the NPT abstainers all along, has also prided itself on being a spokesman and leader of nonaligned nations. Because of pride alone, the Indian delegation was not enthusiastic about following a Swedish lead. When Sweden moved to sign the treaty just seven weeks after it was formally offered for signature, the partnership of interests was even more constricted. Stockholm's decision to sign the NPT could not have been automatic. As mentioned above, the success of the ENDC delegation was incomplete and perhaps not totally obvious. Any leverage over countries such as India might be lost in signing; so also might any leverage aimed at securing further concessions from the great powers.

Yet by summer of 1968 the superpowers had in effect narrowed the choice for countries such as Sweden, as the opening of the treaty for signature on 1 July seemingly precluded any further amendments. For Sweden to refuse to sign would now suggest failure to the home constituency, without promise of imminently redeeming successes, and permanent abstention from the NPT system, with the commercial hardships this might mean. More than any action to date, it would have eased the path for other potential abstainers. After a suitable delay suggesting some disapproval, Sweden signed the treaty on 19 August and prepared to ratify in the autumn of 1968. Stockholm at this point had reason to believe that its signature and ratification would be matched by Italy and Switzerland and that West Germany would not be able to prolong its abstention much into 1969.

The Russians, however, invaded Czechoslovakia in 21 August, upsetting the NPT timetable in many parts of the world. Had Sweden planned to postpone signature even a week more, there might thus have been a delay of many months, with interesting effects on Stockholm's relationship to the pro- and anti-NPT coalitions. As it was, Swedish *ratification* was postponed, as well as Italian and Swiss signature of the treaty; American ratification was suspended (and thereby all ratifications in the Soviet bloc), thus easing the pressure for an early ratification in Stockholm. Yet any very prolonged delay in Swedish ratification was still unlikely; Swedes are not the kind of people to sign treaties without meaning to ratify them. Sweden now would ratify the NPT on 9 January 1970, after both the United States and the Soviet Union had completed their domestic ratification processes. The Scandinavian examples of Norway and Denmark, which ratified much earlier, were thus not matched.

Explicit parliamentary opposition was not expected on the ratification question. The opposition parties were essentially caught off guard by the development of Sweden's ENDC image. What was essentially a multipartisan approach to foreign policy became an asset of the ruling Socialist Party; for example, posters in the 1968 election showed Foreign Minister Nilsson and Mrs. Myrdal together "working for peace." Any upset of the NPT applecart by Sweden now would appear as unseemly as a suggestion of Swedish nuclear weapons, and the unseemly is politically far more self-defeating in Sweden than in many other states. Such opposition or disgruntlement as exists is largely to be found in the Conservative Party, the only party to take an explicit stand in favor of nuclear weapons in the late 1950s. The

Conservatives will now be content to snipe at minor points of the treaty, trying to work within the limits of Sweden's ripened pride in disarmament.

PROBLEMS FOR THE FUTURE

Some problems may yet arise between Sweden and the rest of the world on the exact implications of the NPT. One such issue relates to the research still being conducted by the Swedish armed forces on protection against nuclear weapons, with its possible spillover to nuclear-weapons construction. Such research may soon be approaching a dead end, with little further to be done short of actually beginning construction of bombs. Moreover, such research nominally hinges on the designs of bombs in the arsenals of the nuclear-weapons states; if such states do not acquire new forms of nuclear weapons, because of the test ban or whatever, there will be relatively little further research for Sweden to conduct.

It is nonetheless possible that attention will be drawn to any such research whatsoever as a bad example for other states. Studying construction techniques for the assembly of fissionable materials into atomic weapons does not violate the letter of the NPT, but it can be seen as contrary to the spirit. There are elements in the Swedish government that would like to see an explicit and definite halt put to all such research. Yet other segments of the government, and especially of the military services, would probably protest such a pronounced halt. If the issue were to surface, the Conservatives can similarly be counted upon to make objection.

An outsider could indeed be a little cynical about Sweden's need for such sophisticated research on defense against nuclear weapons. It might be important to know that some bombs produce more blast and less radioactivity and to know the design aspects of such alternatives; yet unless the aggressor advertises which model of bomb he is dropping, the civil defense implications will be less than obvious. Most threatened nations will be content to tell their soldiers to "take cover and pray" if the enemy is about to fire a nuclear warhead; do Swedish soldiers really need to know which design the warhead utilizes?

One might suspect that this research inexplicitly still represents the portion of a military option that the more weapons-minded Swedish elite has been able to sustain. For the moment those who seek to stop this research will have the burden of raising a public issue; with the defensive rationale the program has been dressed in, it is not so clear that it will be stopped.

Defenders of this research note that the prestige of Mrs. Myrdal's delegation at the ENDC has stemmed from a military-scientific expertise that other nonaligned nations do not possess, in part derived from the above "protection against weapons" research. A few such persons even argue that Sweden should have maintained a more explicit weapons program into the mid 1960s, so as to have something to trade the United States and the Soviet Union in exchange for more pronounced concessions.

Aside from such non-bomb-producing military nuclear research (which under the NPT would not in fact be illegal or subject to IAEA safeguards), other issues may also emerge between Sweden and the superpowers, such as how narrowly to interpret NPT restrictions on sales of equipment to a nonsignatory such as India. If a country refuses to sign the NPT or to accept IAEA safeguards and moves to make explosives—peaceful or otherwise—utilizing its own uranium, is an NPT-signatory required to withhold every last bit of equipment that could possibly be used in processing such materials? The treaty refers to equipment "especially designed" for the processing of such materials; this can generate a long or a short embargo list. The U.S. AEC trigger list of such items is generally longer than would be endorsed by Sweden and various other aspiring nuclear equipment suppliers, an issue which the IAEA has not resolved.

Much will depend, as always, on the publicity such sales receive in Sweden. If India quietly puts out tenders for various components of an unsafeguarded reactor, Swedish industry and government may choose to delivery the goods for the market opportunities thus offered, in the belief that another country (such as France, Canada, or Germany) would otherwise make the sale. If an enterprising newspaperman describes the sale as contributing to Indian nuclear weapons, however, the public outcry might force a ban on such sales.

The nuclear question is thus quite analogous to Swedish manufacture of advanced military jet aircraft. Sales of such aircraft abroad strengthen the domestic economic and military potential but possibly violate Sweden's political principles. Some sales are clearly not to be allowed, while other are just as clearly legitimate. In between there exists a grey area in which SAAB can sell airplanes as long as Sveriges Radio does not pointedly comment on it.

However sympathetic Swedish public opinion may be to India, sales to support an explicit weapons program would probably be out of the question. If India convincingly packaged its efforts into what were called "peaceful explosives," some of this reaction might be diffused. Probably nothing could diffuse the reaction to any nuclear cooperation with South Africa. The tolerances of Swedish public opinion and of the superpowers may thus be in or out of step. Where they are out of step the United States and the Soviet Union will be unhappy with the way Sweden has chosen to interpret the NPT.

SWEDEN AND THE BOMB

Swedish political style is characterized by a high regard for domestic and international legality. If one asks a Swedish physicist how long it would take to be able to produce a bomb in 1976, almost invariably his answer ("about six years") will indicate that he presupposes full respect for Sweden's legal obligations; safeguarded plutonium or enriched uranium in Sweden would not be stolen or seized. An Israeli or Indian would be much more prone to discuss the physical possibilities irrespective of treaty obligations.

In this context, therefore, Sweden would for the moment have to begin almost from scratch on any bomb program, since almost all of her reactors would be dependent on an outside supply of enriched uranium; a "legal" and autonomous military program might indeed take five or six years to reach fruition. Sweden's "failure" to develop her military option after 1960 thus essentially refers to her abandonment of natural-uranium reactors which could be fueled domestically for more profitable enriched-uranium reactors. If Sweden were today to violate all of her international commitments, illegally seizing safeguarded plutonium or enriched uranium on her territory, she would be closer to a bomb than ever. Only in the context of its unusual respect for international law has Sweden's military option "atrophied."

Over time it may become commercially appropriate for Sweden to begin to develop its own uranium reserves and then perhaps to construct its own enrichment plant for plutonium, or even a centrifuge-type separation plant for uranium. There are Swedish officials who clearly look forward to the day when the legal gap between Sweden and the bomb will be narrowed again. For them, the vetoing of an agreement with France for development of new extraction methods for Swedish uranium (on the grounds that France would not submit her use of the uranium to IAEA safeguards) was a setback, but "no one is going to move the uranium out of Sweden," and it will become available sooner or later.

Self-sufficiency would do more than reopen bomb options, for certain advantages may exist to make such self-sufficiency generally attractive throughout Sweden. Firstly, such a natural-uranium cycle may someday be more economical than the enriched-uranium cycle the Americans are always peddling; if foreign exchange ever becomes a problem, there will be advantages to indigenous uranium in any event. Secondly, Swedes have bitter memories of being blockaded and cut off from essential imports in two world wars in which they took no part. If Stockholm's electricity ever becomes dependent on the shipment of a replacement reactor charge, limited wars might inflict unnecessary hardship. At present, nations are not allowed to stockpile a large number of backup fuel charges; a large pileup of enriched uranium presents proliferation risks which donor nations may not wish to accept. For Sweden, self-sufficiency in nuclear fuel and electrical power production may thus seem the most comforting solution.

The precise legality of the Swedish political style paradoxically reduces the significance of the NPT relative to other preexisting international agreements. The NPT strictly adds three months to the minimum time period for any Swedish move to nuclear weapons, namely, the three months' notice that must be delivered before withdrawal from the treaty. At the moment other treaty commitments keep this time period at much more than three months, specifically the commitments to foreign suppliers of enriched uranium and of nuclear equipment. As long as such agreements with foreign countries are economically preferable to indigenous development, Sweden will remain legally quite far from nuclear weapons. By 1980, however, much of the Swedish nuclear cycle may again be entirely domestic (according to the 1959 plan it would have been entirely domestic by 1968), so that only the three-month NPT delay (plus some few months more of "crash program" bomb-production time) will remain.

Over time, the spirit rather than the letter of the NPT may indeed make withdrawal unthinkable in Sweden and in other countries. For the moment, many knowledgeable Swedes insist that they were more bound in 1969, before the NPT had gone into effect, than they will be in 1979 under the treaty.

For any country, one can speculatively identify circumstances that would cause a withdrawal from the NPT system or even the launching of a bomb program. For Sweden, however, the circumstances will have to be more than usually speculative. Soviet invasion and occupation of Finland might be such a circumstance. A breakup of NATO, perhaps to be replaced by an integrated Scandinavian political and military structure, might set up the economies of scale to make nuclear arms appropriate again. A drastic erosion of the world's disapproval of nuclear arms might have a similar effect; for example, if all the significant near-nuclear states but Sweden and Canada refused to ratify the NPT.

A West German abstention may set Swedes thinking again in terms of national interests and national strategic location rather than international morality. All the more so any successful German move to ignore the treaty and to acquire nuclear weapons. Perhaps also the simple invidious comparison that would follow an Indian plus an Israeli nuclear-explosives decision (not to mention a Brazilian one) would lead Sweden back into a probomb rather than an antibomb inclination.

Attempts by the Conservatives and others to rechannel Swedish "peace" inclinations toward a reassertion of the nuclear-weapons option will probably be only marginally successful. If most of the near-nuclear states sign and ratify the treaty, even a total absence of great-power disarmament over the next five years will not make it that easy for Sweden to withdraw and go it alone. Since the

NPT and the nuclear question have become so much a question of image for Sweden, her general deference to appearances and proprieties in the policy debate has intensified. Opponents of the NPT and what it represents will have to grumble quietly and discreetly and wait for an issue to develop that autonomously seems to cast doubt on the treaty's value.

8

Brasília, Buenos Aires, Santiago

Brazil has made it clear that she will not sign the NPT.[1] Partially because of this, Argentina and Chile also probably will not sign. All three states signed the Treaty of Tlatelolco, which defines Latin America as a nuclearfree zone,[2] but that treaty, at Brazilian insistence, specifically allows the acquisition and use of peaceful nuclear explosives; the substance of the NPT thus would not be provided by the Latin American treaty.

Various explanations can be offered for the Brazilian stand. The Brazilian government indeed insists that it expects great results from peaceful nuclear explosions in redirecting rivers and opening canals. Citing the overlap between nuclear-explosives technology and other peaceful nuclear technology, the government argues that to give up the first would be to hold back the second.[3] A broader spin-off argument is that Brazil has a brain-drain problem because it cannot find stimulating and rewarding employment for physi-

[1] A comprehensive account of Brazilian attitudes toward the NPT is to be found in H. Jon Rosenbaum and Glenn Cooper, "Brazil and the Nuclear Non-Proliferation Treaty," *International Affairs* 46 (January 1970): 74-90. See also R. Narayanan, "Brazil's Policy Toward Nuclear Disarmament," *The Institute for Defence Studies and Analyses Journal* 3 (October 1970): 178-91.

[2] The text of the Treaty of Tlatelolco (Latin American Nuclear Free Zone Treaty) is to be found in United States Arms Control and Disarmament Agency, *Documents on Disarmament, 1967* (Washington, D.C.: U.S. Government Printing Office, 1969), pp. 69-83 (hereinafter cited as *Documents on Disarmament, 19—*). For related documents and analysis, see Alfonso Garcia Robles, ed., *The Denuclearization of Latin America* (New York: Carnegie Endowment for International Peace, 1967).

[3] For definitive Brazilian statements, see *Documents on Disarmament, 1967*, pp. 135-40, 142-43, 226, and *Documents on Disarmament, 1968*, pp. 49-57, 64-65, 278-83.

cists and engineers tempted to go abroad; a substantial nuclear research program might be the catalyst for the return of such scientists, and invitations to come home, specifically connected to development of a nuclear explosive, have been reported.

DOMESTIC POLITICAL NEEDS

Open opposition to an NPT and affection for nuclear explosives largely emerged during the military regime which took power in 1964, and more precisely, after General Costa e Silva succeeded General Castelo Branco as head of the military government in 1966. Previous regimes had tended rather to stress disarmament as an open foreign policy goal. Brazil, for example, took a strong supporting position at earlier stages of negotiation on the Treaty of Tlatelolco; but in 1964 it began to express reservations even about the effect the treaty might have in keeping American nuclear weapons out of the Panama Canal Zone or Puerto Rico. Mexico favored a Latin American treaty that would bind any states that ratified it; Brazil now insisted that such a treaty have no binding effect until every Latin American republic (Cuba included) had ratified it and until all colonial and nuclear powers outside the Latin American region had ratified certain additional covenants as well. While Mexico and some fourteen other states waived these requirements to make the Treaty of Tlatelolco binding on themselves, the treaty currently has no binding effect on Brazil, Argentina, or Chile.

One can try to explain Brazil's stands in terms of the personal preferences of those dominating the government. While Brazil cannot openly look forward to its first explosive being a "bomb," its air force, army, and navy officers can still contemplate it as such among themselves, positively supporting their *machismo* and self-esteem (but unfortunately perhaps whetting or reinforcing similar appetites in Argentina or Chile). The assertion of Brazilian independence on nuclear matters also satisfies personal aspirations of some important officials in Brazil's Foreign Ministry (the Itamaraty). The foreign service has long been characterized by a high degree of professionalism and competence; as might be expected, such professional self-confidence typically seeks more active foreign policy problems and ventures to which to apply itself. The first army leadership after the coup deposing Goulart was under the control of relatively worldly and pro-United States Brazilian officers who had served in Italy in World War II, staff officer types who saw international politics in terms of the cold war. With the succession to office of General Costa e Silva, the foreign policy tone became more that of the nationalistic barracks officer, quite ready to defy United States lead on any question affecting Brazilian interests, and professionals in the Foreign Ministry thereby regained part of the influence they had lost in 1964. One victim, of course, was the NPT.

Resistance to the treaty does not merely serve the personal preferences of those within the regime. The military government generally has difficulty in capturing popular support and in overcoming its repressive impact in the academic community. A nuclear energy program may help in both areas, impressing the public with Brazil's prowess and creating employment for scientists.

Government-sponsored polls of public opinion, albeit designed to guide respondents toward answers approving government policy, nonetheless demonstrate considerable public awareness and support of a policy of nuclear independence.[4] As many as thirty percent implicitly endorsed even military uses of nuclear energy, although the government has always claimed that weapons were not in question, that any explosives produced would be peaceful. Only about three percent in effect endorsed submission to the NPT system.

There would not of course actually have to be a Brazilian nuclear-explosives program for the rejection of the NPT to make sense politically. Ever since it became a member of the ENDC, Brazil's involvement in world questions of armament and disarmament has been enhanced ipso facto. The world now cares about Brazil's "position," and Brazil therefore has had to have one. If the position was strongly to support disarmament under Goulart, it would be to assert national prerogatives under the military regime. It could therefore make sense for Brazil to insist on retaining options which it had no prospect of ever using, simply

---

[4] The polls are quoted in Rosenbaum and Cooper, "Brazil and the Nuclear Non-Proliferation Treaty," pp. 82-83.

for the domestic and international prestige that would be accrued thereby, prestige which might, but certainly does not have to, become self-validating. If Brazilians and outsiders assume that Brazil must have some nuclear potential because of its recalcitrance on the NPT, so much the better. Such an assumption might lead Brazilian scientists to come home and might even lead them then to generate some nuclear potential; if none of this happened, the regime would still be ahead of the game politically.

Yet the Brazilian government is not interested only in its reputation for scientific skill and national independence; a considerable priority attaches to real industrialization and expansion of the economy. If defiance of the United States on the NPT had seemed to stand in the way of acquiring nuclear electrical power facilities in the 1970s, the Ministry of Mines and Energy might well have been able to overrule the prestige considerations of the Foreign Ministry. The ability of the United States to retaliate when its will is defied in Latin America has been considerably curtailed; for example, a buyer's market is developing wherein any American refusal to sell reactors would quickly be followed by a sale by Canada, West Germany, or Sweden. Brazilian moves toward scientific progress and accomplishment therefore do not require much sacrifice of national dignity and prestige. Indeed, so far as scientific morale is concerned, the two may reinforce each other.

We expect scientists to be socialists and internationalists, worldly in their politics and committed almost instinctively to disarmament. Yet this pattern does not hold for scientists of all nationalities or for all generations within a nationality. Even if the scientific community in Brazil has consistently been more radical or left than the military regimes, a leftism exists in Latin America which is more populist than internationalist, almost instinctively opposed to Yankee imperialism or great-power domination. The government's stand on the NPT, as those on the two-hundred-mile limit and the United Nations Conference on Trade and Development (UNCTAD) negotiations, thus strikes a more sympathetic chord among the Brazilian intellectual and scientific community than outside observers might have hoped. The Soviet Union's strong support for the NPT cuts no ice with the Brazilian military government, of course; but it also has little influence on the radical opposition.

When Brazil was still under the civilian government of Goulart, a government publicly committed to disarmament and to the Latin American Nuclear Free Zone, nuclear energy development was headed by Marcelo D'Amy. While D'Amy held the close confidence of Goulart, he nonetheless favored the fullest independent development of Brazil's nuclear potential, including such potential as would be relevant to explosive devices.

PHYSICAL REALITIES

If Brazil seriously treasures the possibility of peaceful nuclear explosions, no one has yet definitively proven that all such possibilities are silly or without value. The U.S. AEC has been an important source of optimism for such Plowshare projects for opening harbors and canals, and Australians have also taken these prospects quite seriously. It is at least possible that the United States would drag its heels on legitimate projects merely out of fear of proliferation. American private firms also might someday collect what Brazil or another country regarded as excessive profits for services on such a peaceful explosion, even if the U.S. government had acquiesced.[5] If the Brazilian government therefore never approaches nuclear-explosives capability, some move might still be made in this direction simply to force the United States to be reasonable when a legitimate project appears. Leverage emerges with potential even if the potential is not yet exercised. Brazil can reach for such potential without committing itself to more.

To be sure, the reality of Brazilian nuclear potential to date has not been overwhelming. The research reactors in operation all have been supplied by the United States. Extensive reserves of uranium have not yet been uncovered within Brazil; and while quantities of thorium exist, it is generally conceded that an advanced level of nuclear expertise would be required to harness the thorium-U-233

---

[5] The fear that the United States or another nuclear-weapons state would require excessive payments for Plowshare nuclear explosions is outlined in H. A. Keller, H. Bolliger, and P. B. Kalff, "On the Economic Implications of the Proposed Nonproliferation Treaty," *Revue du Droit International* (1968), pp. 44-47.

cycle. Even if the bulk of trained physicists and engineers who have gone abroad were to return to Brazil, there would be a shortage of trained manpower.

Brazil's first power reactor will in fact be an American model purchased from Westinghouse, fueled with enriched uranium obtained from the U.S. AEC. The choice reflects a decision to place the economic advantages of this form of electrical power production ahead of whatever political independence or military options might accrue from a natural-uranium reactor. Argentina two years earlier had made an opposite choice. Thus the choice of reactor here does not suggest an early move toward peaceful nuclear explosives. The world scientific consensus has for the moment turned more pessimistic about the practicality of peaceful explosions; currently no one in Brazil is seriously agitating for any crash program for such explosives.

Yet these days a lack of indigenous capabilities does not necessarily end all hopes of political independence. Brazil can accept American equipment and IAEA safeguards on a project-by-project basis, at the same time accumulating useful experience and consolidating cadres of Brazilian and foreign scientists. If projects for the moment stress economic sensibility ahead of the production of explosives, the country will remain legally free to launch a more indigenous program whenever it chooses to make the "peaceful bomb." If Brazil at some later point acquires a natural-fuel reactor which is free of American services and IAEA safeguards, there might be little to prevent the hiring of engineers from Germany or some other country to assist in running it to produce plutonium as well as electricity and to configure such plutonium into a crude explosive. Once it has been ratified, the NPT will forbid Germany to supply fuel or equipment for such purposes. The treaty cannot however prevent individual scientists from migrating from country to country, perhaps to acquire experience with explosives denied them in their own homeland. Charges are even circulated from time to time in Eastern Europe that Bonn and Brazil will thus conspire to allow Germans to produce nuclear explosives (bombs) in Brazil. An agreement also exists for nuclear cooperation between India and Brazil, and it would certainly not be impossible for India to produce explosives within the decade. The Brazilian government, by the very nature of its vehement claims on the legitimacy of peaceful explosives, is incapable of swearing that it will never be party to anyone else's development of such devices.

The search continues for usable deposits of uranium within Brazil. Even if such deposits are slow to appear, natural uranium will be available from abroad, or perhaps even enriched uranium if the American monopoly is finally broken by one or another of the groups hoping to enter the enrichment business. Uranium might be available from Portugal; Brazil and Portugal have always been close in culture and history, and political relations have improved considerably since 1964. The pariahs of the world tend to cluster together. If Brazil elected to cut itself off from the NPT community, Portugal has already been ostracized in many ways and indeed has not yet chosen to commit itself to the NPT. Even more ostracized and more abundantly endowed with uranium, of course, is South Africa, a partner of Portugal's in resisting black threats to the African "white redoubt" and ideologically sympathetic to the anti-Communist regime in Brasília.

Brazil's slowness in developing a nuclear industry might even have become an advantage in terms of political independence. A nation more advanced on nuclear matters would already have purchased larger power reactors from the United States and would thus have developed more of a vested interest in continued access to the enriched uranium with which to fuel them. American assistance to Brazil has to date included only small research reactors. Brazil indeed does not have a very pressing need for nuclear power. It will be well into the 1980s before all the potential sites for hydroelectric power have been exploited. Brazil therefore has not needed to worry as much as some about whether its resistance to the NPT would bring retaliatory withholding of technological assistance by the United States. Such retaliation has not materialized for any of the countries which have defied American wishes on the treaty; yet it has still been easier for Brazil than for Japan to risk such a response.

BRAZIL'S NEIGHBORS

Since all the significant near-nuclear states in the Southern Hemisphere have expressed some disquiet about the treaty, we might look at the problem in "hemispheric" terms. With regard to nuclear testing, such a focus indeed makes sense meteorologically, since wind patterns tend to dump each hemisphere's radioactivity on its own side of the equator. Yet comparisons with Australia or South Africa play a very small role in Brazilian deliberations on proliferation and the NPT. Each of the ABC powers—Argentina, Brazil, and Chile—is mostly interested in comparisons with the other two. A rejection of the treaty by any of these three will tend to prejudice chances of signature and ratification by the other two. Bomb manufacture by one would suggest similar moves by the other two.

Argentina is ahead of Brazil in the field of nuclear energy, even if one disregards President Peron's claims in the early 1950s that a nuclear-weapons program was underway. The number of trained physicists is greater, and the work done on research reactors has been more extensive. Perhaps more importantly for the long term, Argentina has deposits of uranium, while nothing definitely extractable has been identified in Brazil. Argentina has moreover purchased the first powerproducing (and therefore plutonium-producing) reactor in South America,[6] from the Siemens concern in West Germany; the reactor (unlike those installed in West Germany) is designed to utilize natural instead of enriched uranium as a fuel and thus to by-pass the fuelprocessing services of the U.S. AEC. The reactor has nonetheless been placed under IAEA safeguards for the present, perhaps because Siemens insisted on this, if not also because the reactor will require heavy water as a moderating agent, which currently must be obtained from the U.S. AEC.

Brazil, perhaps because of its earlier membership in the ENDC, has been allowed preeminence in explicating South American opposition to the NPT. Argentina's diplomatic style has indeed been much less eloquent and explicit than that of its neighbor; Argentina has been more cautious and reserved, simply failing to ratify the Tlatelolco Treaty rather than being "bound to be bound" once all the superpowers accept the protocols, indeed never getting around to ratifying the Test-Ban Treaty. Yet the Argentine government, in the person of President Ongania, has echoed and endorsed the Brazilian objections,[7] and Argentina has

---

[6] *New York Times,* October 27, 1969, p. 19.
[7] Ongania's statement of April 1969 is cited in Rosenbaum and Cooper, "Brazil and the Nuclear Non-Proliferation Treaty," p. 84.

not signed the NPT. If the interrelationship of the three ABC countries is characterized by rivalries of prestige, politics, and even military strength, this has not precluded cooperation in denouncing and resisting the treaty.

Hints have appeared also of broader technical cooperation between Argentina and Brazil, without details on the activities in question.[8] Argentina could certainly supply natural uranium and engineering expertise to the Brazilians. If the Argentine scientific complex were not large enough to be efficient on nuclear matters, Brazil could in exchange offer to combine its staffs of physicists and engineers with those of Argentina, in hopes of more speedy nuclear progress for both countries. Yet the very nature of Latin American nationalism makes much pooling of effort unlikely. Each nation is eager for industrialization and prosperity, but all the factions in each are eager also for national independence and autonomy; therefore it is likely that duplicate nuclear programs, rather than any significant merger, will be pursued.

A merger would have one enormous advantage beyond any economic savings it might generate. Just as the Euratom structure has reassured Belgium and Germany about each other, a full program of Argentine-Brazilian exchange might automatically assure each state that the other was not producing nuclear explosives and/or bombs. If IAEA coverage under the NPT will not fully cover the ABC countries, it is not so easy to see what will preclude the possibility of a nuclear arms race in the 1980s; it may be all too possible for rumors to appear in Argentina about Brazilian bombs or vice versa, triggering precautionary responses and in the end giving the continent two or more nuclear-weapons states when none wanted to acquire weapons in the first place.

Skeptics on such scenarios could point to the recently good relations between Argentina and Brazil and to the relative infrequency of wars in South America. Relations between each of these countries and Chile have been less good since the advent of Allende, but Chile is really a small country, having barely 9 million people, with very little at all as yet underway in the nuclear field. Yet pessimists can point to the earlier race of Argentina and Brazil to match each other in useless aircraft carriers; where weapons as deadly as atomic bombs are concerned, it is not impossible that the mere threat of such weapons production could drive nations

[8] Ibid., pp. 88-89.

apart, even when there were no purely political issues at hand. Internal struggles for power in Argentina and domestic disquiet in Brazil have outweighed thoughts of the NPT or of nuclear explosives in both countries. Yet neither is likely to lose track of the issues or to be content with doing much less than the other. Cooperation and rivalry on nuclear matters can thus drive each nation forward. A more radical regime in Chile may yet renounce nuclear-weapons options as a matter of high moral principle, regardless of what Brazil and Argentina do; on the other hand, it may instead choose to emulate Castro's rejection of the NPT and the Treaty of Tlatelolco. If Brazil were indeed to detonate a nuclear explosive, the Argentinian and Chilean regimes would be under strong pressure to match it, no matter what the domestic problems and no matter what the ideological tones of the regimes.

9

London, Ottawa, Canberra

The English-speaking world has not seemed a source of trouble for the NPT, unless one counts India and South Africa as "English-speaking countries." We normally apply the term to the United States, the United Kingdom, and the predominantly white members of the British Commonwealth (Canada, New Zealand, and Australia). The United Kingdom, like the United States, seemingly has no reason to "make trouble" for the treaty, since its birthright to nuclear weapons is preserved by the NPT; but as we shall see, this somewhat oversimplifies British national interest. Canada and New Zealand have behaved themselves quite acceptably by endorsing the view that a halt to proliferation is urgent. Indeed New Zealand signed the NPT on the day it was offered and ratified it in October 1969; Canada ratified even earlier, in January 1969. By a somewhat misleading projection, Australia also has escaped suspicion about its resistance to the treaty, although suspicion might be quite appropriate.

LONDON

Almost until the treaty was presented for signature, the British government seemed determined to behave exactly as the United States and the Soviet Union, as if its privileged status as a nuclear-weapons state required logically that it do so. If anything,

British statements in favor of the treaty in Geneva were more doctrinaire than those of the two superpowers. Lord Chalfont, the British delegate to the ENDC, acquired a reputation for his unremitting personal belief in the rightness of the treaty; at times he became acrimoniously entangled with the representatives of the near-nuclear states which were allegedly conspiring among themselves to gang up on the treaty. This style reflected several negative as well as positive factors. There obviously was some embarrassment at the extent to which the United States and the Soviet Union had together privately composed the treaty, consulting Great Britain no more than anyone else. Being favored by a treaty on which it has not been asked to help, Britain could not loudly object; thus it might have felt driven loudly to applaud. Apart from this, Great Britain's prestige interest would be served by reminding the world that it was a nuclear power and entitled always to remain one under the NPT. As long as one's own nuclear birthright is not threatened, one naturally "has an interest" in halting proliferation.

Great Britain might also have been relieved to escape any embarrassing requests for direct or indirect assistance on nuclear weapons. If India were now to desire some particular form of equipment, IAEA safeguards would be required. If Australia felt that the British nuclear testing on Australian soil in the 1950s had established an obligation to supply nuclear warheads on request, the NPT would finally terminate such expectations.

There were thus practical as well as psychological arguments for Great Britain resolutely to endorse the treaty in Geneva and elsewhere. Despite the Soviet invasion of Czechoslovakia, Britain indeed rushed to ratify the treaty on 27 November 1968, thus becoming the third state to ratify (after Ireland, which was ever antiproliferation, and civil-war-torn Nigeria) and the *first* nuclear-weapons state.

At some point, however, the realization had to emerge that Great Britain's position might be more special, given that she was suing for entry into Europe, which includes Euratom with its objections, real or imagined, to the NPT. If Germany is to be the principal partisan of British entry into the European communities, it will simply not do for Great Britain to act as if it regarded Germany as the "nth" country.

Perhaps the first sign of change in the British position came with the replacement of Lord Chalfont as the delegate to the ENDC; at the same time the Indian government moved Ambassador Trivedi from Geneva to Vienna in what has been described as a mutual deescalation of the pro- and anti-NPT sides. More serious tokens since then have included British willingness to participate with the Netherlands and West Germany in a consortium to develop centrifuge processes for the enrichment of uranium.[1]

The centrifuge issue carries with it a number of symbolisms. The United States has always been reluctant to see even discussions of such processes, for fear that such discussions will someday lead to the simplification of the production of weapons-grade uranium to the point that any country could produce A- or H-bombs. Critics have accused the U.S. AEC of using world-wide fears of nuclear proliferation to maintain its near monopoly on the enrichment of uranium for use as reactor fuel, a monopoly which perhaps has kept prices higher than they should have been and which in any event has constricted the potential independence of nations now dependent on the U.S. fuel supply. Conversely, those who are skeptical of the economic practicality of centrifuge processes accuse European and Japanese enthusiasts of trying to pressure the AEC or of trying to establish a credible posture of going it alone in case their resistance to the NPT would cause them to be cut off from the U.S. fuel supply. The move to pool British, Dutch, and German technology would thus be a political or business gesture rather than a serious scientific effort. If so, it still finds Britain placing its weight on the opposite side of the negotiating scale, strengthening rather than weakening Germany's ability to be slow in accepting American advice or IAEA jurisdiction.

German participation in the consortium has inevitably drawn attention on both sides of the iron curtain. Currently only two plants are planned—one in Great Britain and one in the Netherlands—obviously a concession to Russian and other sensibilities. British and Dutch delegates could presumably outvote any German move toward bomb production within the decisionmaking process of the consortium itself. Yet Germany is not in the mood to go after bombs for any time in the foreseeable future. Political

[1] For information on the Centrifuge consortium, see John Maddox, "The Nuclear Club," *Survival* XI (September 1969), pp. 275-80, and articles on "Non-proliferation" in *Survival* XI (March 1969), pp. 74-78.

victory for Bonn comes whenever someone concedes or testifies that Bonn is to be trusted; British willingness to consort with the Germans on uranium enrichment is thus a positive political gesture, enhancing Bonn's reputation even as Moscow's propaganda has sought to run it down.

Britain redefined its role again early in 1970 with proposals for revisions and compromises in IAEA safeguards practices, now that the mandate for such safeguards would come under the NPT rather than bilateral commercial arrangements. The British proposals here were obviously intended to capture the role of honest-broker middleman between the IAEA, the United States, and the Soviet Union on one hand and the Euratom countries on the other.

OTTAWA

It is sometimes said, in cruel jest, that Canada will be the first country to give away nuclear weapons without making any of its own. Canadian opposition to acquiring nuclear weapons has been remarkably consistent; a country that could have produced such weapons as early as 1955, it has come to interpret abstention as a sign of independence. Canadian assistance to Indian nuclear development, on the other hand, was of great technological significance, and the legal constraints on the exploitation of such aid were less clear than they could have been.[2]

Canada's willingness to forswear nuclear weapons must be attributed largely to its political and geographic closeness to the United States.[3] It is almost inconceivable that an enemy could attack Canada without suffering retaliation from the strategic forces of the United States. If other links were not enough, Canadian population centers are all so close to the border that their destruction would be physically felt in the United States. The treaty relationships which bind the two countries are really

---

[2] A full description of Canadian cooperation with India may be found in Leonard Beaton, *Must the Bomb Spread?* (Hammondsworth, Middlesex: Penguin, 1966), pp. 71-77.

[3] A comprehensive account of Canadian views on the NPT can be found in Michael Sherman, *Nuclear Proliferation: The Treaty and After* (Ottawa: The Canadian Institute of International Affairs, 1968).

nothing more than formalizations of a symbiotic relationship which has been obvious since the 1930s. Even a mere threat of nuclear attack against Canada would draw U.S. reactions little different from those to a threat against the United States itself.

Under such circumstances there are very few plausible Gallois-type scenarios wherein the United States would fail to protect Canada and an indigenous deterrent to keep Canada from being invaded or destroyed would then be necessary. Moreover, politically there is no inducement to seek such scenarios; rightly or wrongly, Canadians have seen the American government as wanting to thrust nuclear weapons upon them rather than withhold them. If one shows his political independence in France or India by reaching for the bomb, Canadians show theirs by turning it down.

This attitude, fortuitous as it is for the NPT, has some interesting historical roots. When the Eisenhower administration was trying to maintain the legitimacy of tactical nuclear weapons in the face of Russian efforts to delegitimate them, it was not above encouraging Canadian units in Germany and France to participate in maneuvers predicated on such weapons. Canadian liberals wishing to show moral superiority over U.S. Republicans were thus antibomb. The U.S. offensive deterrent forces which would destroy the Soviet Union in retaliation would have to overfly or land in Canada, presumably with bombs on board. Later air defense systems might use nuclear warheads, basing supporting equipment or even the warheads themselves in Canada. Despite President Kennedy's deemphasis of nuclear options after 1961, disputes on the acceptability of Bomarc air defense system warheads caused an open rift between Kennedy and Prime Minister Diefenbaker.[4]

If Canadians in 1968 had seen America as many other "nth" country publics do, namely, as selfishly trying to monopolize nuclear weapons for itself, they might have asserted a very different style of independence. But public attitudes change relatively slowly. Until the very last minute, Canadian support for a halt to proliferation was almost unequivocal, because responsible governmental officials indeed were concerned about proliferation

---

[4] See Peter C. Newman, *Renegade in Power: The Diefenbaker Years* (Toronto: McClelland and Stewart, 1963), pp. 340-54, for an account of the various issues between Kennedy and Diefenbaker on nuclear weapons.

and also because the man on the street sees a compatibility rather than a conflict between renouncing nuclear weapons and staying independent of the United States.

Mexican support for a Latin American nuclearfree zone and for the NPT reflects the same neighbor-of-the-United States syndrome: namely, that if an independent brandishing of such weapons is ludicrous in the shadow of such an enormous neighbor, one can perhaps show independence and moral superiority by making a fetish of renouncing such weapons. It has been a long time since any American proposed basing nuclear warheads on bases in Mexico; and the Mexican government would take great delight in rejecting such a request.

It would be unfair to characterize Canadian attitudes toward the NPT solely or even primarily in terms of psychological independence of the United States. Nuclear proliferation is a serious problem for the world and for Canada; Canada has good reasons—not only psychological ones—to forgo bombs and to urge other nations to forgo them. Yet Canada also may have some reasons to question the NPT, since the treaty inevitably involves more than just preventing nuclear weapons spread. By the end of the NPT debates, Canadian representatives had indeed begun to express some sympathy for other near-nuclear states' objections to the treaty, and Canadian signature was affixed not on the first possible day (1 July 1968), as had been expected, but a few weeks later on 23 July. To the extent that general acceptance of the treaty would have been aided by a "stampede effect," as was the case with the Test-Ban Treaty, Canada's delay did not contribute to the stampede.

It would not be totally uncharitable to attribute some of Canada's change in tone to the realization that the United States was wholeheartedly behind the NPT. In no sense is one now spiting the United States by signing the treaty; if any part of the Canadian public or government had begun to sense this new psychological reality, it might have become suddenly appropriate to express some reservations. More sensibly, Canadian influence among other near-nuclear countries, especially India and Pakistan, would not have been enhanced by a singleminded endorsement of the treaty. It certainly made sense for Canada to admit that some objections to the NPT were well-taken in principle. The delay in signature moreover was apparently due to the coincident change in government in Ottawa, with Trudeau becoming Prime Minister; for a new head of government to review policy before signing an important treaty is certainly understandable.

Although Canada is averse to bombs, it is not averse to all forms of nuclear industry. It is in Canada's interest to market its large reserves of uranium and to develop peaceful uses of it for domestic and foreign application. It is difficult to work with electricity-producing reactors without having some larger market within which to recover the costs of research and development; Canadian sales to India thus served commercial purposes and allowed Canada to indulge its impulse to offer assistance to an economically less developed country. Since other nations have sought the same markets, little criticism would be forthcoming on the sales themselves. There has been some criticism, however, of the controls Canada imposed to ensure that fuel and reactors would be used only for peaceful purposes.

About Russian nuclear assistance to Communist and non-Communist countries we know little except that relatively loose and ineffective safeguards systems have been instituted. We do know that France sold Israel a reactor and that the sale was accompanied by no public pledges for peaceful use and no visible controls. With these very significant exceptions, Canada's first arrangement with India was the laxest known, calling for no inspection of the Canadian-Indian reactor at Trombay and pledging only that India would use the materials for "peaceful purposes."[5] Whether "peaceful purposes" include peaceful nuclear explosives (the "peaceful bomb") is thus extremely debatable, with India ready to insist that it does and Canada not at all resolutely opposed. We will have to wait to see whether India will argue further that nuclear weapons used for deterrence serve only a "peaceful purpose."

One can be too critical of Canada's first arrangements. The United States was itself demonstrating flexibility on safeguards questions at the time (for example, it decided to give a special exemption to Euratom). It could easily be argued then—and might be today—that a soft and pliant approach would maintain one's influence without antagonizing public opinion in the countries involved. A gentlemen's agreement was what the Ottawa-New

---

[5] See Beaton, *Must the Bomb Spread?*, pp. 71-72.

Delhi agreement most resembled, and gentlemen's agreements are often more effective than overly long and detailed contracts. Through the 1950s, moreover, Canada was still earning considerable sums by supplying uranium to the United States for military weapons production. The drift of this form of sale did not encourage excessive moral introspection on the ultimate purposes of fuel delivered elsewhere.

In the 1960s conditions changed to make Canada considerably more strict about such sales of fuel and equipment. Such sales can still be economically important and thus can still occasion concern that other nations might win away markets by more permissive policies. But cuts in American production of nuclear warheads in the early 1960s produced some economic distress in the Canadian uranium-mining industry, which serves as a sobering antidote to any temptation for enormous sales to "nth" country markets. Resigned to a generally less booming but perhaps more stable market for uranium, Canada has now settled on a policy of sales only for peaceful purposes, only under safeguards. French requests for uranium in 1965 thus were rejected when Paris would not accept IAEA safeguards.[6] American requests for new raw materials for weapons are only hypothetical for the moment but presumably would have to be rejected. Indian requests for additional assistance according to the earlier pattern have also been refused in the wake of protests from Pakistan and the general consensus that tighter guarantees of nonexplosive use are needed.[7] In 1965 agreement was reached for a reciprocal inspection of the Rajasthan reactor by Canadians and of an identical reactor at Douglas Point, Ontario, by Indians. This concession to inspection was indeed accepted only reluctantly by India. While Canadian surveillance of the reactor at Rajasthan has been reasonably careful, Indian inspection of Douglas Point has been most perfunctory, implicitly conforming to the Indian argument that rigorous safeguards are not necessary among nations which have basic political trust and confidence in each other.

For the longer run, the NPT is very desirable from Canada's viewpoint, if only because it makes IAEA inspection standard for all sales by Canada or anyone else. The costs of inspection can get to be burdensome, not only for the nation being inspected but

[6] Ibid., pp. 33-34.
[7] Sherman, *Nuclear Proliferation*, p. 76.

also for the nation executing safeguards, especially if that nation is exporting significant quantities of fuel. The antagonism generated in negotiating safeguards arrangements also can be troublesome; thrusting this responsibility onto an international body may thus make it easier for Canada to maintain a friendly climate with nations such as India. It was indeed not beyond the Canadian imagination in 1955 that India might someday seek to divert its nuclear assistance to purposes which Canada and the world might not approve. Yet friendly relations and the completion of transactions often enough require taking risks; and other sellers of nuclear expertise might have undertaken such risks if Canada had been unwilling.

There are still other possible causes for Canadian disquiet about the NPT. Canada after all is a country which has proved itself almost as much as any nation could, having resisted the temptation for more than fifteen years. The NPT will require safeguards on any materials Canada sells to countries presumably less trustworthy, but it will also impose inspection within Canada itself. As is argued by almost everyone, inspection can be costly and troublesome; the Canadian electrical power industry will thus not be enthusiastic about such inspection. Given the record, one could assume that IAEA inspectors would tend to be easygoing in Canada, but they can still waste some time and resources. An interesting question is whether the NPT requires that such inspection extend even to mining; the relevant documents hardly settle this issue one way or the other, and South Africa has cited the possibility of such mine inspection as one reason to hold back signature. For the moment the Canadian government is prepared to argue that mining should not fall under safeguards, for it faces considerable internal objection from the mining industry if Vienna's mandate should be interpreted as requiring them.

CANBERRA

Australia has had to endure relatively little publicity about its stand on the NPT. Until West Germany signed the treaty, Australia, in sharp contrast to Great Britain, Canada, and New Zealand, could quietly leave the pact untouched. After Bonn and

Tokyo signed, Canberra on 27 February 1970 did the same but stated that ratification was by no means decided. Ratification was to be postponed for some time, perhaps forever, as the world watched and waited for the equally unassured German and Japanese ratifications.[8]

There has been no agitated public discussion of the NPT in Australia. From the outset the opposition Labour Party tended to speak favorably of it, while the governing Liberal-Country coalition seemed reserved. A group of Australian writers on arms control urged early signature as a positive contribution to the halting of the world arms race.[9] Prominent Australian nuclear physicists, on the other hand, as quickly declared their opposition to the treaty, arguing that development of Australian nuclear industry would be significantly frustrated by the terms of the NPT.[10]

In principle, the renunciation of nuclear weapons might constrict peaceful nuclear research and development in the nonexplosive categories, but this is far from proven for any of the near-nuclear states. More directly, Australia, like Brazil, has professed an interest in the peaceful use of Plowshare nuclear explosives, especially for dredging new harbors. If signed, the NPT would require that such explosives be obtained from one of the nuclear-weapons states, which Australia might find unfairly constricting. For example, it seemed at the beginning of 1969 that the U.S. AEC was eager to help Australia to dredge a harbor by means of a nuclear explosion at Cape Keraudren in Western Australia.[11] For whatever reason—because Canberra had not yet signed the NPT or because Australian plans for the project were badly thought out—American interest very quickly cooled, and the project has not moved ahead. Australians may have been toying with a peaceful explosion just to set a precedent; yet the

---

[8] A good general discussion on Australian attitudes on bombs and the NPT can be found in Ian Bellany, "Australia's Nuclear Policy," *India Quarterly* 25 (October/December 1969): 374-84.

[9] See Hedley Bull, "In Support of the Non-Proliferation Treaty," *Quadrant* 12 (May/June 1968): 25-29, for a strong statement in favor of Australia accepting the NPT.

[10] See "Australian Doubts on the Treaty," *Quadrant* 12 (May/June 1968): 30-34.

[11] See the U.S. AEC memorandum on the Cape Keraudren project printed in U.S. Senate, Ninety-first Congress, First Session, Foreign Relations Committee, *Hearings on the Non-Proliferation Treaty* (Washington, D.C.: U.S. Government Printing Office, 1969), pp. 329-30.

precedent set was that in the future such explosions would depend heavily on the goodwill of the oligopoly the NPT establishes.

The Australian Atomic Energy Commission (AAEC) was quite slow to invest in research reactors or to commit itself to power-producing reactors, even though countries such as Great Britain had already somewhat prematurely decided that nuclear energy would be the prime source of electricity for the future. Given the repeated disappointments in the cost-effectiveness of nuclear electricity as compared with fossil fuel, the slowness of Australian commitment here need hardly be lamented in terms of civilian economic returns. The impact for Canberra's political and military options is more mixed, however. The failure to move to nuclear power delayed Australian acquisition of valuable engineering experience in the nuclear field; but it also averted the installation of reactors, which would have made part of Australia's electrical production continually dependent on American good will.

The leadership of the AAEC thus hardly did all it could to terminate suspicions or hopes that Australia might someday move to make its own atomic bombs.[12] Australia's earliest research reactors were indeed fueled by enriched uranium, which must be obtained from the United States. When faced with a choice between enriched-uranium and natural-uranium electricity-producing reactors, however, the AAEC for a time urged the latter; cost comparisons, although somewhat debatable, tend to suggest that the former is more efficient. AAEC spokesmen defended such a choice as enabling Australia to avoid future fuel-supply problems, that is, dependence on the U.S. AEC for enrichment of Australian or other uranium for use as reactor fuel. "Fuel-supply fears" can refer to the difficulties of shipment across oceans or to the dangers of being asked exorbitant prices. They can also pertain to the leverage of fuel suppliers, who might use their fuel supplies to force an earlier acceptance and ratification of the NPT or to prevent an Australian decision to produce plutonium bombs indigenously.

By 1972, the Australian choice seemed to be veering toward an enriched-uranium design, but only after it became more likely that

[12] A detailed discussion of Australian choices in the nuclear energy area can be found in Harry Gelber, *The Australian-American Alliance* (Hammondsworth, Middlesex: Penguin, 1968), pp. 39-55.

new processes for uranium enrichment could be developed in various places around the world to break the American monopoly. Australian reactors of either variety would of course depend at first on technical assistance from Britain, Canada, or the United States. IAEA safeguards would thus be imposed whether or not Australia ratified the NPT. At a later stage, however, domestically-developed reactors might be fueled with unsafeguarded fuel, as long as the NPT had not yet bound Australia.

It is always somewhat insulting to speculate about the motives of the nuclear scientific bureaucracy of a nation that has taken a stand against the NPT. If reluctance is plausibly phrased in terms of civilian electrical power or peaceful explosions, can one politely charge that lust after weapons, or at least after weapons options, also plays a role? Yet can anyone deny that the status and funding of an Atomic Energy Agency are likely to be enhanced when its public is ready to consider a nuclear-weapons program?

Australia was thus close to becoming a country in which a small group of nuclear physicists could physically prepare a de facto nuclear-weapons option and veto a legal renunciation of such weapons. If the country's political climate had remained favorable or even indifferent, this scientific bureaucracy would probably have determined policy. As it was, only a major political turn against retaining nuclear-weapons options could upset this.

Have ordinary Australians shared a craving for nuclear weapons and national power and prestige? Have they had real enemies to fear? Has it been easy for political parties to win elections by showing interest in the nuclear-weapons option? It is still too early for many voices in Australia to have been raised in favor of an explicit Australian nuclear-weapons program; only the militantly anti-Communist left-of-center Democratic Labour Party has ever endorsed weapons explicitly. Yet strong political support has been at hand (paradoxically even among some NPT backers) for preparatory moves to lay the technological groundwork for any hypothetical bomb decision.

There has long been a discussion in Australia, on an abstract level, of a Southern Hemispheric policy on nuclear questions,[13] originating with proposals from Labour spokesmen that the Southern Hemisphere be made a nuclearfree zone into which no

---

[13] See Bellany, "Australia's Nuclear Policy."

nuclear weapons whatsoever would be deployed. The hemispheric focus could easily enough have become reversed over time, however, since South Africa has not yet signed the NPT and since Brazil has declared it will not sign (and therefore also Argentina and Chile probably will not adhere to it). Australian cooperation with the nuclear program of Great Britain, including the provision of sites for atomic bomb tests, historically has made some Australians feel that it would only be natural for this weapons technology to be shared. If Japan did not ratify the NPT, moreover, many Australians might well also consider not ratifying. Australians on both the left and right still share a certain amount of distrust of Japanese military power.

A comparison of Australian and Japanese attitudes on the NPT has not had to be phrased totally in the competitive terms of neither trusting the other. If either government had any non-military qualms about committing itself to the treaty, it welcomed the company of the other in the position of nonratification. Even if no progress were ever made in drafting a Japanese-Australian nuclear community to resemble Euratom and thereby to win any special treatment accorded the latter, it still would have been possible for the two countries to consult informally on how to fend off pressure for ratification or how to extract concessions in "uni-national" dealings with the IAEA. When physicists and diplomats know each other as well as they do on the NPT question, coordination of antitreaty positions can be so tacit and informal as to preclude any charges of formal collusion and conspiracy.

Yet Australian official statements in 1968 and 1969 did not focus on fears of Japan in excusing a retention of the nuclear-weapons option; rather they focused much more consistently on alleged fears of Communist China, which after all had its atomic arsenal already in hand. This was consistent with the general anti-Communist ideological tone of Australian foreign policy and presumably made it easy to reconcile delay on the NPT with friendship for the United States. Few Australians could really claim to be worried about Chinese Communist (or Russian) nuclear strikes directly on Australian cities. The United States has still seemed committed to guaranteeing Canberra against any such attacks, as the identification of Americans with Australians, which has been strong ever since the shared experiences of World War II, had been formalized by mutual defense treaties. A blatant atomic attack on Sydney would thus still very plausibly have brought American retaliation against the attacking country.

A U.S. nuclear umbrella for defending Australia against any likely form of conventional attack was inherently less plausible, however. If Indonesia, for example, had again engaged Commonwealth and Australian forces in Malaysia or North Borneo, there would have been no likelihood of American nuclear forces being deployed in any way helpful to Australia; similarly, if there had been an insurrection supported by Indonesia or China in eastern New Guinea.

Events of the last four years in Indonesia most assuredly reduced the plausibility of Communist control there, but given the general volatility of Indonesian politics, the possibility of drastic reversals could never be excluded. (One need only remember that President Sukarno was hinting at an Indonesian nuclear-weapons program only a few years ago,[14] perhaps with the implication that Peking was assisting Indonesia in the production of such weapons.) Regardless of whether such claims had any reality, they made it easier for Australians to discover some lingering value in their own nuclear option when considering possible threats to themselves from Southeast Asia. If it made any difference at all, Indonesia did not sign the NPT until 2 March 1970, just after Australia, and has not ratified.

From the viewpoint of the Liberal-Country Australian Government of 1968, moreover, the Soviet-American agreement on the NPT was most saliently accompanied by Lyndon Johnson's announcement that he would not run for reelection and that the bombing of North Vietnam would be terminated, which implied a reduced American commitment in the Far East. However feigned or exaggerated Australian fears of Chinese Communist guerrilla expansionism may have been, a reduced U.S. interest in the Far East was an important change, and these events precluded any Australian haste in endorsing the NPT.

Yet the tone of Australian policy was not to remain frozen in the anti-Communist style of the Liberal-Country coalition. Domestic dissatisfactions may have made it inevitable that the

---

[14] See *New York Times*, July 25, 1965, p. 17, for rumors and reports on Indonesia working toward a bomb.

Australian Labour Party would win power at some time in the early 1970s. While foreign policy issues were probably not central to the Labour victory, changes in the external environment at least reduced foreign policy as an asset for the incumbents. The defeat of the Communists in Indonesia indeed made it much more difficult to convince Australian voters that they were threatened by guerrilla aggressive island-hopping from the north. The frustrations of the American intervention in Vietnam, with its token participation by Australian forces, similarly took the edge off any enthusiasm for allied cooperation with the United States. Talk of Australia "playing a role as a power" had emerged in the late 1960s in response to American calls for assistance in containing the alleged Chinese Communist menace; such talk now seemed far less appropriate to international politics, such that Labour could phrase some of its appeal in terms of Australia "setting a good example" and "aligning itself with the third world." Since resistance to the NPT had been enmeshed in the anti-Communist rhetoric, such popular resistance was thus doomed to weaken somewhat as the rhetoric changed.

Ratification of the NPT in January 1973 was not without some costs to the Labour government elected a month earlier. The bureaucracy of the AAEC remained quietly opposed to the treaty (and one always loses something by alienating such bureaucracies). Japan has not yet ratified the treaty and might someday move to acquire nuclear weapons; will the Australian electorate of that time, or its government, regret having given up the weapons option?

Yet the new regime was in the mood to make its foreign policy changes quickly, ranging from recognition of Communist China and East Germany to the withdrawal of the Australian contingent from Vietnam. The Labour government characterizes itself as less weapons-minded and more prodisarmament than its predecessor; thus ratification of a treaty that had already been signed seemed only the most natural step toward responsible and cooperative international behavior. Such a ratification incidentally at last brings Australia into line with New Zealand, which has also just elected a Labour government. Both the new governments have hinted that they might consider interfering with French nuclear tests in the South Pacific, thus showing a disposition to reach out to their public on the nuclear-weapons issue.

The public in Australia will be more enthusiastic about protesting French detonations than about renouncing Australian weapons options; the NPT ratification thus can not be seen as any direct concession to public opinion. Compared with the antitreaty attitudes of leftists elsewhere, for example, the Japanese Socialist Party and the Chinese Communists, the NPT may not even be so "left." Having composed a foreign policy image which included renouncing nuclear weapons, however, the Labour Party chose to live with it after winning, whether or not it will help at the polls in the future.

Many observers would originally have expected Australia to duplicate Canada's or New Zealand's speedy signature of the treaty, on the grounds that nuclear weapons are inappropriate for such small countries, on the assumption that geopolitical or cultural ties presumably have assured British or American support in any crisis. But the nuclear-energy elite of Australia resisted the NPT more than other Commonwealth members, and the political elite tended to see more of a military defense problem. For its public, moreover, Australia for a time may have been a plausible Far Eastern "power," more so, obviously, than New Zealand.

The predictions of Australian support for the treaty at long last are being confirmed, but some damage may yet have been done to the NPT during the years that Australia remained the only English-speaking country resisting the treaty. Australian ratification will now focus increased pressure on Japan to ratify; but the pressure might come too late.

10

Bonn

West Germany finally signed the Nuclear Non-Proliferation Treaty on 28 November 1969. When the treaty is ratified in Bonn, the Federal Republic will no longer have the right to manufacture its own nuclear weapons or to accept them from any foreign donor. Equally significant, it will agree to accept the safeguards inspection controls of the IAEA over its peaceful nuclear activities in addition to or in place of the Euratom controls which presently exist.

West German signature is indeed significant, both for Germany and for the treaty.[1] Foreign observers had feared that Bonn might delay signature much longer or even explicitly reject the treaty; such a delay or rejection in turn might have triggered enough other rejections to make the treaty an obvious failure.

One could have begun this discussion of the NPT with West Germany; almost everyone else begins here. There are indeed

[1] Statements of German views on the NPT are to be found in Theo Sommer, "The Objectives of Germany," in *A World of Nuclear Powers?*, ed. Alastair Buchan (Englewood Cliffs: Prentice-Hall, 1966), pp. 39-54. See also articles by Theo Sommer and Carl F. von Weizsacker in "National Viewpoints," *Survival* IX (May 1967), pp. 144-49. A 1967 interview with Minister for Scientific Research Gerhard Stoltenberg illustrating the strongly stated anti-NPT case can be found in *Réalités Magazine* (English ed.) 203 (October 1967): 27-29. See Stephen D. Kertesz, ed., *Nuclear Non-Proliferation* (South Bend: University of Notre Dame Press, 1967), for a reprint of a parliamentary statement of Willi Brandt quite evenhandedly discussing German reactions to the NPT. See also German official statements in United States Arms Control and Disarmament Agency, *Documents on Disarmament, 1967*, pp. 48-54, 90-92, 92-96, 106-7, 160-62, 179-82, 206-16, 623-25, and *Documents on Disarmament, 1968*, pp. 152-55, 488-90 (hereinafter cited as *Documents on Disarmament, 19—*).

Americans within the Department of State who agree that German signature is the major object of the treaty; according to this view, for historical and other reasons Germany should be the last to have the psychological burden of possessing nuclear weapons and the suspicions that would accompany it. A more common American view has been that Germany is far from the most likely to be an "nth" power, being committed after all under the Western European Union Treaty not to produce nuclear weapons. Since Bonn is the Soviet Union's number one object, however, Bonn might also be at the top of the United States list, since nothing can be accomplished on the NPT unless the two superpowers stick together. A third view would dismiss as having no absolute significance Moscow's warped view of the future of international politics but would nonetheless place Bonn at the head of the list simply because so many other nations have been hiding in West Germany's shadow. Once Bonn signed the treaty, attention could turn away from West Germany to focus again on Israel, Japan, Australia, and India.

On reflection, a West German signature was always more likely than more pessimistic observers would have thought; the signature was largely a matter of timing, a question not of whether but of when. The fact that Germany signed may not, in the end, at all disprove the wisdom of those responsible for delay, however, for Bonn's procrastination induced or was accompanied by some important changes in the environment erected by the Soviet Union and the United States. West Germany may have had to sign the NPT; but she signed at the right moment.

THE SOVIET UNION

West Germany's confrontation with the Soviet Union has probably been the most serious cause for delay. Spokesmen of the Auswärtiges Amt in Bonn enumerated long lists of concessions demanded of Moscow, if the Soviet Union was serious about the substance of the treaty. West Germany already stood committed to the substance of the NPT vis-à-vis its Western allies, the argument ran, and it was then only a question of allowing the Soviet Union to become party to this West German renunciation of nuclear weapons. If the Russians desired this, they had to offer something in exchange.

At earlier stages, West Germans were prone to argue that the nuclear-weapons option should only be surrendered in exchange for substantial progress toward German reunification. The imminence of the NPT and other events of the past five years indeed made this a less realistic request, and the "intervention" in Czechoslovakia saw West Germany's concern for the sovereignty of its own territory take precedence over its thoughts of challenging Pankow's sovereignty. Even before Czechoslovakia, Bonn and Moscow had become entangled in an exchange of statements on the relevance of Articles 53 and 107 of the U.N. Charter, which Moscow interpreted as sanctioning its own armed intervention in West Germany to prevent a resurgence of Nazism. Explicit demands for a renunciation of these articles may have induced an explicit reassertion of them, or vice versa. Yet at the least it would have been possible for Moscow to begin stressing Article 2 more than Articles 53 and 107. When "intervention" was still in the air, West Germans easily could depict the NPT as just one more excuse for Soviet aggression under these articles and could urge a delay in signature until the Soviet tone was altered.

In discussing the NPT itself, Russian spokesmen continually implied that West Germany was the only state where the world had to fear further nuclear proliferation. It was difficult to get the Soviet Union to admit that it feared such weapons in anyone else's hands; for example, Russians would tell New Delhi that her signature was only necessary as a means to increase pressure on Germany to sign.

As long as the Soviet Union had not yet ratified the NPT, therefore, the interpretation would arise that such ratification was conditional on German signature and that the NPT was specially designed and needed in order to control West Germany. A German signature in 1968 or early 1969 would seemingly have been an admission of the validity of this interpretation, and the German Foreign Ministry indeed withheld its signature until Soviet ratification was assured.

To watch for specific quid pro quos with the Soviet Union may thus have been a little misleading; the general tone of Bonn's relations with Moscow was more important. If the Russians felt that they must continually slander and threaten Bonn to keep the

East German regime viable, Bonn would have to be reluctant to sign an NPT. A thawing of relations with Moscow conversely made the treaty much easier to sign.

The relationship between Bonn and Moscow has never been entirely hostile, despite the black picture that officials in the Auswärtiges Amt have been prone to paint, but it also has not been very predictable. In some situations, the essence of politics is to exaggerate disagreements. It seems clear that Moscow feared and hated Bonn less than she pretended. It is also likely that Bonn feared and distrusted the Soviet Union less than it claimed. While German Foreign Ministry spokesmen regularly imputed hundreds of ingenious and devious intentions concerning the NPT to Moscow, relatively sophisticated discussions also occurred whereby Bonn and Moscow sounded each other out on steps toward détente. It would be a mistake to assume that Germany really was as opposed to the NPT, because of distrust of Soviet intentions, as it claimed to be; a case of trouble on the Sino-Soviet frontier, for instance, could suddenly occasion a visible warming in relations between Bonn and Moscow or Bonn and Warsaw. Trade particularly makes cooperation of some sort in the interests of these states; yet public images can take on a life of their own. Whenever Bonn and Moscow claimed to distrust each other, a treaty which seemed to have been dictated from Moscow could win little genuine acceptance among the German public.

In Bonn and elsewhere, an interesting debate occurred as to whether the Soviet Union really preferred that West Germany sign the NPT. It was at least possible that the Soviet Union did not imminently fear German nuclear weapons but rather feared the obvious appeal of the Federal Republic when compared to that of the East German régime. If the DDR inherently must always lack popular support, Moscow perhaps must regularly work to reduce Bonn's prestige; a West German failure to sign the NPT might just accomplish this. In this view, the Russians had to keep the treaty just obnoxious enough to drive Bonn to reject it, which could explain Russian statements about the NPT and about the U.N. Charter.

Yet the Russians obviously were not totally indifferent to whether West Germany obtained its own nuclear weapons. It is true that there are already some 7,000 American weapons on

German soil, but these are not quite the same. The Russians also are not oblivious to the success or failure of the treaty outside of Europe; if Germany could kill the treaty by not signing, the Soviet Union preferred that it sign.

It is also not clear that German prestige indeed would have suffered so quickly or so much from a refusal to sign the NPT. The righteousness of the treaty is in doubt; and, by asserting its independence, Bonn might have won far more friends in New Delhi than it lost in New York. Under such circumstances the NPT would have become a very weak stick with which to beat Bonn over the head; thus the argument for wanting Bonn to sign became all the clearer. Soviet leaders indeed seem to have accepted the argument that obvious Russian pressure would be counter-productive for the treaty, at least during the 1969 German election campaign; Moscow remained relatively quiet about the NPT during the summer of 1969, and the treaty did not become a campaign issue in Bonn.

THE IAEA

Bonn has identified its other principal bargaining adversary (that is, other than the Soviet Union) as the International Atomic Energy Agency (IAEA) in Vienna, the safeguards agent under the terms of the NPT. The exact way in which the IAEA will exercise its role remained to be specified, even though the treaty text can no longer be changed, and many were convinced in Bonn that German signature, or at least ratification, should be withheld until these specifications were conceded in Germany's favor.

The issue most often cited by Bonn was Vienna's threat to duplicate Euratom's inspection system, thus making it redundant. Bonn's ideal would have been a complete exemption of Euratom from IAEA safeguards, since the United States had granted it such an exemption in the 1950s (on the assumption that the French and Belgians could be counted upon to preclude any German manufacture of weapons). The maximum fear was of an IAEA procedure exactly parallel to that of countries with no preexisting international inspectorate at all. Hopefully a compromise could be found to ease the German inspection burden and to keep the institutional image of Euratom alive. If delay in signing extracted some prospect of compromise, it may thus have been advisable.

There are Germans who remain sincerely committed to the goals and symbols of European unity. European "nationalism" since 1949 has indeed been a safety-valve alternative to a resurgence of German nationalism; a setback to European institutionalization will thus have some costs. One can argue that Euratom has been unimportant compared with the European Coal and Steel Community (ECSC) and the European Economic Community (EEC); indeed, Euratom's only real function has been precisely the bookkeeping controls which the IAEA threatens to preempt and make redundant. Yet the issue had been raised openly in Germany, so that a total surrender of Euratom's special prerogatives now might seem just one more admission that European unification is no longer likely.

Euratom's positive ventures in nuclear industry have been beset by arguments on three basic issues: (1) how much to do at the community level, (2) in which nations to locate installations, and (3) what kind of equipment to develop. On each of these, the six member nations have aligned themselves differently.

Larger nations, such as France and West Germany, have seen advantages in stressing their national programs, while the Benelux nations have preferred an enhanced Euratom program in which they might participate and exploit economies of scale. On the question of where to locate installations, every member nation naturally has wanted to channel into its own territory as much capital and employment opportunity from such projects as possible. On the choice of kinds of equipment, France has tended to stress the use of natural-uranium-fueled reactors so as to avoid a dependence on the fuel-enrichment facilities of the United States; the other member states, however, have preferred the greater profitability of reactors fueled with enriched uranium. Thus, France typically has been outvoted on this issue and therefore has become quite bitter about the fact that Europe's political independence of Washington was sacrificed simply for a greater profit margin.

While the last issue is the one most relevant to the NPT, the interaction of all three has kept Euratom from reaching the levels of positive activity once projected for it, with the result that the "negative" function of controlling nuclear materials has become the principle justification for its existence. Although it is most often Germany that voices concern for Euratom, Belgians and Italians are also concerned,[2] and perhaps even Frenchmen, albeit very little pro or con, is said about the treaty in Paris. Italians have expressed a concern that the treaty will doom Europe by permanently enshrining Great Britain and France as nuclear-weapons states and relegating other states to a secondary status, thus upsetting the symmetry and parity from which union presumably would evolve.

German, Italian, and Belgian objections to the IAEA have not stemmed only from concern for European unity. There is a definite fear that Vienna's safeguards program will be much more costly than Euratom's has been, mainly to those being inspected. Many real or imagined reasons for such fears can be offered: the inexperience of Vienna as compared with Brussels, the relative indifference (or venality) of outsiders inspecting plants within Europe, the deliberate troublemaking by inspectors from Communist countries, or the likely bureaucratic imperialism of an agency like the IAEA. At the least, it must be recognized that Euratom is more widely known among German and other industrialists; if their ideal was no inspection at all (and therefore maximum profits), Euratom has not been an unbearable second choice.

Euratom's (or Germany's) suit for special treatment has inevitably faced an obstacle of invidious comparison from other advanced nuclear countries, especially Japan. Japan has had to have a two-front position, favoring all possible relaxations of Vienna's overall inspection coverage as long as Japan shares each of such relaxations with Euratom, but strongly opposed to any concessions in which Japan would not share.

For the time being, one can argue that Euratom indeed offers the world adversary mechanisms that the Japanese national system

[2] An official Italian allusion to conflict between the NPT and European unity can be found in the statement of Foreign Minister Fanfani to the ENDC, reprinted in *Documents on Disarmament, 1967,* pp. 312-15. A clear exposition of French and other national attitudes on Euratom activities can be found in Robert Gilpin, *France in the Age of the Scientific State* (Princeton: Princeton University Press, 1968), pp. 403-13.

can never offer; surely French inspectors can be counted upon not to acquiesce in a clandestine German bomb project. Yet the Soviet Union has consistently expressed an unwillingness to trust this kind of adversary relationship, and other states have echoed the notion that Euratom's inspectorate would not be sufficiently credible. Moreover, if Euratom is to lead to unification, the adversary relationship ultimately will have to terminate. Bonn could not have it both ways; if Euratom was to be defended as a stepping stone to European unity and to an end of European mistrust, Bonn could not cite this mistrust as a reassurance to the outside world that no atomic bombs would be produced in 1979.

Because of Euratom's greater experience with monitoring an entire system, its monitoring might have allowed for better controls at lower disturbance to what was being controlled. Yet the likely growth of plutonium flows in the next decade may render all prior experience trivial or irrelevant, in both Brussels and Vienna. To do as perfect a control job in as inobtrusive a manner, when ten times as much material must be accounted for, may challenge Brussels's competence also. German physicists working on safeguards techniques indeed are quite open-minded about whether Euratom has any advantages over the IAEA on a technical basis.

The same physicists have become quite influential within the IAEA itself. There is hope that the need for direct access by human inspectors can be reduced by the introduction of "black-box" automatic measuring devices, on which a German research team at Karlsruhe had done considerable investigation.[3] It is even possible that Germany can earn profits by producing and selling these devices to the IAEA for use around the world. Whichever was more significant, avoidance of inspection burdens or sales of devices for avoiding inspection burdens, some issues remained to be talked over before Germany gave away all its cards. Bonn has played its cards quite well by placing a number of its scientists into strong positions for shaping IAEA safeguards arrangements.

---

[3] For illustration of the sophistication on safeguards practices and philosophy emerging from the research facility at Karlsruhe, see W. Hafele, "Safeguards on Nuclear Materials," in *Impact of New Technologies on the Arms Race*, ed. B. T. Feld et al. (Cambridge: The M.I.T. Press, 1971), pp. 237-49.

THE UNITED STATES

The Soviet Union and the IAEA aside, Bonn hoped for concessions from its American ally and indeed obtained some. The German requirements for American concessions here are less concrete, reflecting rather the tone of American policy toward NATO and détente with the Soviet Union. Germany still depends significantly on American commitments to the defense of the Federal Republic and Berlin and perhaps to an ultimate unification of Germany, and it values periodic reassurance of these commitments. Private dealings between the United States and the Soviet Union, whatever the substance, do not reassure Bonn; in 1965 Germany did not feel that it had been adequately consulted on the development of the NPT.

Since 1949, Bonn has generally seen its own security as dependent on the presence of American troops and nuclear weapons in Germany and on American distrust of the Soviet Union. These alone guarantee against the experience of the Czechs in 1948 or 1967. But financial pressures continually threaten to reduce the commitment of United States troops to West Germany. Arms-control fears of nuclear weaponry similarly suggest a removal of American nuclear weapons and threaten to launch a more general détente in which the United States and the Soviet Union would reach agreements without consulting Bonn. If this is an irrational nightmare, the Soviet-American negotiations on the NPT in Geneva nevertheless bore a disturbing superficial resemblance to it.

The Nixon administration came into office generally unenthusiastic about the treaty, more committed to consulting European allies on all matters of policy. This change in style went a long way toward satisfying German requirements, but questions remained, for example, on how the U.S. government would interpret its obligations if an impasse should arise in the IAEA-Euratom negotiations. Delaying a German signature while the new American administration settled into a durable policy groove thus also seemed quite sensible in Bonn.

Delay has produced some improvements from Bonn's point of

view. In the process of U.S. Senate hearings on ratification, West Germany obtained at least a few verbal concessions favoring its security position. First, it was noted that any termination of the NATO treaty in the American view would constitute the "extraordinary events" allowing a signatory to give three months' notice of its withdrawal from the NPT.[4] An American government twenty years from now certainly will not be bound by the statements of any secretary of state in 1968 or 1969, but such statements on the record nonetheless would tend to make the United States more sympathetic, rather than less, if Bonn then must move to nuclear weapons.

Over the shorter term, the interpretation was advanced that wars of certain sorts would ipso facto terminate the legal obligations of the NPT.[5] Not every war would terminate the treaty, to be sure. All-out war between the superpowers indeed would terminate it, while a small local war involving only non-nuclear-weapons powers would not. Whether a Soviet ground assault on West Germany would be considered as belonging to the former or the latter type is left suitably unclear. We thus have an interpretation that the ban on proliferation could be lifted as soon as a major war with the Soviet Union had broken out in Europe, an interpretation giving Bonn at least some of the psychological weight it had been seeking, since thousands of American nuclear weapons already had been deployed within reach of West German military forces trained to use them.

The timing of the final activation of the NPT illustrates the perennial delicacy of East-West relations, and especially of relations between Bonn and Moscow. When President Nixon, after some two months in office, decided to go ahead with a request for Senate ratification of the NPT, the West German government feared that Moscow would make its own ratification explicitly contingent on German signature; wishing very much to avoid the implications of having to bow to Russian pressure and *diktat,* Bonn apparently persuaded Nixon to withhold deposit of ratifica-

---

[4] U.S. Senate, Ninety-first Congress, First Session, Foreign Relations Committee, *Hearings on the Non-Proliferation Treaty* (Washington, D.C.: U.S. Government Printing Office, 1969), p. 424.

[5] U.S. Senate, Ninetieth Congress, Second Session, Foreign Relations Committee, *Hearings on the Non-Proliferation Treaty* (Washington, D.C.: U.S. Government Printing Office, 1968), p. 43.

tion until a joint and simultaneous Soviet deposit could be assured. Affairs stood almost at a standstill for much of the spring of 1969, except that ratifications began to emerge from satellite regimes—from Poland, Hungary, and Czechoslovakia—at least hinting that the Communist world still wanted the treaty.

President Nixon now pointedly avoided urging NPT signature on Bonn, reasoning that such pressure was counterproductive to winning West German support for the treaty. Washington attempted to persuade the Soviet regime that an immediate ratification by both superpowers could actually accelerate German signature, by removing the implication that either power was pressuring Bonn in conditionally withholding ratification. But the Soviet Union had more than one reason for withholding ratification until Bonn had signed. Moscow's desire to focus embarrassing pressures on West Germany aside, the Soviet leadership could very reasonably fear that premature ratification would be followed by legal reservations on a German signature, reservations crippling the treaty's meaning.

The final compromise thus had to be elaborate in order to protect each side against humiliation by the other. The stage was set in November 1969 with a joint Soviet-American announcement of "completion of all the domestic ratification processes in Moscow and Washington," accompanied by a promise that the ratifications "would be deposited" jointly, soon, and as a matter of course. This announcement of almost-ratification by Moscow was then quickly (almost conspiratorially) followed by German signature; joint deposit of the American and Russian ratifications then came when the number of other non-weapons-state ratifications had exceeded the forty necessary to make the treaty operative. The diplomatic hassle on whether the Russians or Germans were first to give in had been happily obscured.

THE DOMESTIC FACTOR

In terms of German external interests, the delay in signing the NPT may have been either too long or too short. It would be fortuitous if it turned out to be of optimal length. Yet it would be a mistake to sum up the treaty solely in terms of its effects abroad; the impact on German domestic politics is of enormous significance, and delay here may indeed have been optimal.

Public reaction to the treaty has not been extensive. Yet the German public increasingly demands that Bonn's foreign policy be effective rather than simply affected. After a long period in which the Federal Republic necessarily had to ratify the initiatives of its allies, Germans could have turned with zest to any opportunity to not sign or ratify or to even generate initiatives of their own. Such a new assertiveness in a country defeated in World War II will seem to have sinister implications, albeit nationalism is now regarded as quite natural and fitting elsewhere. Japanese opinion, which is passing through similar stages of development, indeed even on the same issue, attracts a little less concerned comment.

Any great disgruntlement to the effect that democratic German foreign policy was again showing itself submissive and ineffectual on the NPT question would have given the nondemocratic fringe parties such as the NPD an issue to exploit in the 1969 elections and later ones. It was important that some indigenous momentum seem to be imparted on behalf of the treaty, rather than simply the momentum of pressure from abroad. The summer of 1969 thus fortuitously provided a climate in which it was much easier for Bonn to accept the treaty, as the election brought to power a coalition of Social Democrats (SPD) and Free Democrats (FDP), parties which had been on record as supporting the NPT for some time; by contrast, the Christian Democrats (CDU) had been lukewarm and their Bavarian affiliate, the Christian Socialists (CSU), was actively opposed to the treaty. The election was close; the particular outcomes could hardly be regarded as pre-determined or inevitable. The SPD drew fewer votes for its party slate than for its individual candidates, and the FDP indeed had lost much of its prior strength. The NPT was not really debated in the election campaign, so that the election's result was not a very explicit mandate for signature.

Clearly the CSU under Franz Josef Strauss reasoned that the NPD must be preempted, with the "nationalist" issue of resistance to the Nuclear Non-Proliferation Treaty being harnessed within the democratic spectrum. Under certain circumstances Strauss might correctly have reasoned that a complete rejection of the American-Soviet initiative was necessary to show that Bonn was a sovereign entity. But the alternative view would stress timing, that it was important that the United States and the Soviet Union cease to seem to be pressuring Germany and that Bonn hold its signature until a moment when it could seem to be self-determined. To sign under Johnson would have been visibly to capitulate to pressure; to sign under Nixon was something very different.

The mandate for the treaty which was generated in the elections was very weak; it is nonetheless possible and desirable for Germans to believe that they did produce such a mandate. The NPT is at its best when nations believe they themselves are choosing its obligations, rather than being forced to accept them. A signature before 1969 might have been tainted with hints of West German submission and ineffectuality in foreign politics. American and Russian shifts to a soft-sell posture in the summer of 1969 thus went a long way toward making the NPT a more promising instrument for the future.

The new mandate is in the interests of the Social Democrats, as well as of the Germans and others as a whole. It can perhaps be molded into a vindication of earlier Socialist anti-weapons positions and a rebuff to earlier CDU stalling. One must still be concerned, however, about whether the pseudo-mandate for the treaty will survive until the time comes for ratification. If the same fears of the treaty as before got a hearing, if superpower pressuring again became obvious, the final swallow could be much more bitter than this initial sip.

THE REAL COSTS OF DELAY

It has been difficult for outsiders to predict how Germany would respond to the NPT; and it has been just as difficult for Germany to predict how the outside world would respond to German procrastination. More pessimistic predictions have indeed been disproven. The United States, Great Britain, Belgium, and other nations have shown little anxiety about German abstention from the treaty. Whatever the far-sighted concern of the arms-control experts of the various powers, the man on the street has not coupled memories of World War II with Germany's debatable nuclear option.

Indeed the publics of most NATO countries are not especially aware of the treaty's existence, or else their feelings are divided. There are also reservations about the treaty in the Benelux nations and in Italy, in part because of what it may do to the Euratom structure. Since Bonn has phrased many of its objections to the NPT in terms of Euratom, a knowledgeable Belgian could not react very negatively to Bonn's stance. If Britain now aspired to join the six (which would mean also joining Euratom), it too could hardly indulge in any accusatory tirades against Bonn on what Bonn describes as Euratom issues.

However much France wished Bonn to sign the NPT, for the moment France itself is committed to decline to sign; thus it is again difficult for French public opinion to wax indignant against Germany. The United States, having elected a president who showed little enthusiasm for the NPT, has also not really gotten worked up about the issue. In all these countries, a prolonged German refusal to sign the treaty might have damaged Bonn's standing, but there is no evidence of damage to date.

Bonn's delays might have cost more than the suspicions of its allies; they might conceivably have increased Russian, Polish, or Czech fears of German revanchism, allowing Ulbricht or someone else to head off any détente with West Germany. Signature of the NPT indeed has been accompanied by a thaw with Eastern European countries, but which has come first is debatable. There will always be some uncertainty as to whether a Communist fear of revanchism was sincere or whether it masked concerns for the viability of the DDR. If it were the latter, such concerns would be overcome more by the attraction of trade with West Germany than by Bonn's second renunciation of nuclear weapons. Election of the SPD, acknowledgment of the Oder-Neisse line, signature of the NPT—to be sure, all served to accelerate such détente as was possible. Yet the NPT may not have been crucial to such détente; indeed the relative warming of relations between Moscow and Bonn had already begun in the spring of 1969.

It is true that, for as long as Bonn had not signed the NPT, what little attention the treaty drew anywhere was heavily focused on West Germany. The signature, happily, may change this, if NPT commentators do not now become fixated on the question of German ratification. Public opinion trends have imposed different costs on Germany from what might have been predicted. The world has not encountered a new wave of Germanophobia; one wave of swastika daubings would have done more harm to the German image than the entire eighteen months of coyness on the treaty. There was indeed no proclivity toward anti-German feelings anywhere outside the NATO and Warsaw Pact areas. Japan, India, Brazil, and Zambia have not brooded about German rejection of the NPT; if anything they have welcomed it. Even Israel has used Bonn's quibbles as an excuse to avoid signature.

Rather the cost for Bonn comes precisely from other states opposed to the treaty escaping publicity in West Germany's shadow, which allows them more time to establish a position for rejecting the treaty. If Germany indeed did not care about nuclear weapons spreading in Asia, Latin America, or the Middle East, then nothing would be lost by delay; however, reasonable Germans care at least a little.

Germany professes to resent the amount of attention devoted to its decision on the NPT. Perhaps a fairer distribution of world press coverage would focus more on Israel, India, Japan, and Australia than on Bonn. Yet there is something self-confirming about this process of assigning attention. Precisely because the world at first thought Germany crucial to the treaty, it has become crucial; other states have deferred their decisions, secure in the assumption that Bonn's stance allowed them to do so.

THE NECESSITY OF THE TREATY

West German signature of the NPT may retrospectively seem to have been inevitable, but one must still specify the real pressures which accounted for this inevitability. One must also try to assess the likelihood of treaty ratification.

As on all of Bonn's other foreign policy questions, United States attitudes have had an overwhelming influence on Bonn's consideration of the NPT. An explicit West German rejection of the treaty, even during the Nixon administration, would have strained the alliance with the United States more than most Germans would find acceptable. President Nixon might never have pushed the NPT in the first place, but the United States stands substantially committed to the treaty by Johnson, and this investment can not simply be thrown away.

Political reactions aside, American adherence to the NPT could have imposed material costs on West Germany if Bonn persisted in rejecting the NPT system. The treaty's text requires that signatories not supply equipment or fuel to non-weapons states except under IAEA safeguards. Since German power reactors depend on the U.S. AEC for enriched uranium as fuel, Bonn would have had to come to terms with Vienna either inside or outside of the treaty. Any prolonged resistance to the treaty couched domestically in terms of the shortcomings of the Vienna Agency could thus have led to a fuel-supply breakdown.

The American position on fuel supply has hardly been made explicit; warnings were never leaked that enriched uranium would become a lever to force signature of the NPT or to force submission to any and all demands of the IAEA. Yet the United States clearly has wanted Bonn to ratify the treaty, and it would not automatically tolerate every German objection to IAEA inspection.

Technically and economically, West Germany has few alternatives. Uranium is potentially available from South Africa, but all of Germany's reactors were designed for enriched uranium (because profitseeking German businessmen found the U.S. AEC's fuel prices cheaper). Reactors for natural uranium have been designed in Germany and sold to countries like Argentina, but none as yet have been installed in Germany. As potential rivals to the American fuel supply, one can turn to the Anglo-Dutch-German consortium on centrifuge enrichment processes or to the existing British and French enrichment plants. Yet the latter are limited in capacity, and the former will not come into commercial operation until the end of the 1970s at best. Thus a completely independent fuel policy is a very extreme solution which might for a time severely constrict Germany's electrical supply.

An explicit rejection of the treaty would moreover encourage speculation about nuclear weapons, and most Germans today do not want such weapons. Such speculation might not have arisen if the NPT had never been proposed. By asking Germany to promise twice, all of the powers have weakened the impact of her having promised once; the NPT signature may have been necessary simply to make Germans feel that they were internationally bound into the 1970s. If delay or rejection of the NPT had become conjoined with a serious discussion of nuclear weapons, some real deterioration might indeed have occurred in German relations with the NATO and Warsaw Pact states. What is excessively hypothetical today would perhaps not always have remained so.

For Bonn, the NPT has thus become a question of ratification. Ratification should not be rushed, if only because haste would seem improper in the face of the many possible drawbacks that Germans have ascribed to the treaty, and more importantly, because all the Euratom countries have been pledged to hold back ratification until Brussels and Vienna have reached agreement on safeguards procedures.

It would be politically dangerous to seem prematurely or too clearly committed to the treaty. Charges of commercial espionage will inevitably be made against IAEA personnel; there will possibly also be incidents where inspectors (perhaps from Communist countries) demand more thorough inspection practices than Germans feel are justified. It may thus be easy for the NPD, the CSU, or others to win electoral points by saying "I told you so." The treaty's benefits for Germany may lie in what it prevents (proliferation abroad, for example) and thus may be relatively invisible; costs will be more visible. If the treaty is thus a necessary evil for Germany and other non-nuclear states, it will not very much enhance the electoral glamour of its supporters. Ratification is indeed likely, but as it has had to be delayed, it might not yet be totally assured. As before, much will depend on how other aspects of German domestic politics and foreign policy now evolve.

The SPD will almost certainly remain firmly committed to the treaty, although the treaty has hardly been the issue on which the rank and file expend all their enthusiasm. Most Germans, including the membership of all the parties except possibly the CSU, indeed have not yet been caught up in the NPT issue. After a seemly delay, enough CDU votes could probably be found to ensure an easy ratification, as the CDU has never been clearly against the treaty.

Internationally, it will be important for the superpowers to continue their soft-sell approach to the treaty and for the world not to become too fixated on German ratification. The new style of President Nixon gave Bonn some crucial room on one side of the international scene at the beginning of 1969, and in the summer the Russians seemed to recognize a need for leeway on their side. The Soviet Union will probably now continue to avoid pressing the Germans because of other prospects for détente and/or because it really wants the NPT to succeed.

Paradoxically the same Russian-German détente that made possible a dignified signature of the NPT also delayed ratification. The détente brought forth a German-Soviet nonaggression treaty and a German-Polish treaty accepting the Oder-Neisse Line, treaties which captured much more German public attention than the NPT and which logically had priority over the NPT in terms of action. Some improvement of the access conditions around West Berlin was indeed made an explicit prerequisite for Bonn's ratification of the Russian nonaggression pact. When negotiations on Berlin were prolonged by the special interests and concerns of the East German regime, the NPT was delayed simply in terms of having to keep its place in line. Public concentration on the Berlin question in any event precluded all sense of urgency on NPT ratification.

11

Rome

Italy's difficulties concerning the NPT serve as a good example of the peculiar difficulties immobilizing NPT opponents in many countries of the world. The legal questions posed by the treaty are hardly trivial; Italy has been a member of Euratom and would thus have to adjust for the first time to IAEA inspection. Yet, while there are potentially significant arguments against Italy accepting the NPT, the opposition has not really succeeded in capturing an audience, and the treaty has been signed.[1] The NPT has been interpreted simply as disarmament or as a "first step" toward it, and the moral strength of this image has precluded more sophisticated levels of discussion, at which a real debate might have emerged.

OPPOSITION TO THE TREATY

There are indeed some interesting arguments against Italy accepting the treaty. Foreign Ministry officials are accustomed to trading concessions for counterconcessions; although there is no sentiment in Italy for producing nuclear weapons, the legal option of producing them might have been held for barter against something in exchange. No clear quid pro quo for Rome emerged.

---

[1] An expression of Italian dissatisfaction with the treaty can be found in Achille Albonetti, "The NPT Draft Under Scrutiny," *Survival* IX (July 1967), pp. 223-26.

The NPT is noted as the first "unequal treaty" of recent times, the first clearly classing signatories as privileged or unprivileged. Along the lines of a similar "national interest" stance, it has been noted that few other Mediterranean states intend to be bound by the treaty: Israel will not sign; the United Arab Republic will not ratify unless Israel signs; Albania will not sign; France will not sign and already has nuclear weapons; Spain and Portugal have not signed; Greece and Yugoslavia delayed their ratifications into 1970; and Turkey has not ratified.

Like other near-nuclear states, Italy potentially can lose some economic gains if the costs of IAEA inspection should prove to be serious. Italian industrialists will echo the claims of Germany and other countries that Euratom's methods are less obtrusive and more effective than those of the IAEA. If one must discount much of this as premature and propagandistic, the methods of Brussels are at least relatively known, and the personnel all the more so. Vienna's methods, which threaten to be new and to be added on top of those of Euratom, could be costly.

Related to the defense of Euratom is the attitude of Germany. Euratom has long been one of Bonn's several reasons for not yet accepting the treaty. An Italian could have argued that German-Italian cooperation here would have been of mutual interest and that Germany's gratitude for Italian nonsignature would have been redeemable at some point. Excessively explicit cooperation might have been counterproductive, but the two nations after all are still NATO allies, as well as comembers of the European Communities.

Apart from the opposition's substantive arguments, one can look at the Italians who belong, however quietly, to the opposition. Most significant of course are the Foreign Ministry career professionals, who for the reasons outlined above see Italian interests as gratuitously compromised in the treaty. Officers of the armed forces generally share this aversion to the NPT, if only simply because nuclear weapons might someday be needed or because any disarmament treaty, any proscription of weapons, reduces the prestige and status of the military (yet the professional soldier, burdened by memories of World War II and Fascism, is not overly influential in Italy today).

In the political spectrum, the Neo-Fascist Italian Social Movement opposes the treaty, although it has not made it a central campaign issue. The opposition of the Neo-Fascists hardly hurts the treaty and perhaps helps it by tainting any nationalist opposition to it. The Liberal Party, which has good connections in the business community, has also criticized the treaty, presumably on the grounds that it threatens commercial profits and Italian economic expansion.

PRESSURES FAVORING THE TREATY

Really effective opposition to the treaty would require a generation of mass popular opposition, through leaks to the press, through relatively demagogic political campaigning, and so on. Yet if this could be managed in Germany or Japan, there are serious obstacles to it in Italy.

To begin, the NPT has not generally been presented as an anti-Italian treaty by any of its foreign sponsors. Russian (and American) spokesmen for the treaty have warned of nuclear weapons in the hands of Germans or perhaps Indians, Israelis, and Japanese, but there has been little recourse to representing Italians as Frankenstein monsters. Also, the NPT question is a relatively esoteric one which does not relate directly to any of the more serious issues besetting Italian politics. The political struggle for the moment hinges on domestic issues, such as the role of the Church or of the Left in Italy's future. Since questions of nuclear proliferation do not easily plug into this basic conflict, it is more difficult to involve the average citizen.

Perhaps the one area in which the treaty hits home is economics. Every Italian presumably should be concerned lest sales and profits be lost due to controls or inspection from outside. Italian businessmen presumably should have been lobbying the public to protect their own profits. Yet the Italian nuclear industry, after an initial spurt which saw Italy become the world's third-ranking state in terms of installed nuclear electrical production, has been in relative doldrums for the past five years. In the face of criticisms of its investment, it has been difficult for the nuclear industry to exert pressure against the treaty.

When compared with the opponents of the treaty, the supporters indeed seem strong. First, there are (at least nominally) all the normal government coalition parties: the Socialists, the Social

Democrats, the Republicans, and the Christian Democrats. For the Socialists and the Republicans, the treaty is a straightforward condemnation of weapons or militarism, an endorsement of international peace. Regardless of the specifics of the treaty, it is a vindication of longstanding positions and thus has to be pushed forward and endorsed; the ideological baggage of any piece of legislation is significant, and the NPT implicitly vindicates the Left.

One might quite properly expect the Christian Democrats to be less enthusiastic about the treaty, which indeed they are. Yet it can hardly be the party of the Pope and still talk openly about a need to reserve weapons options, and the United States is a principal supporter of the treaty to boot. Open opposition to the treaty would have required a more painstaking preparation of the psychological environment, but the Catholic church, which is increasingly committed to peace and deténte, has hardly set this up. If there was any hope that the Church might support Italian right-wing resistance to the treaty, this was of course decisively dispelled when the Holy See acceded to the treaty on 25 February 1971, perhaps primarily as a signal to the larger country which surrounds it.

Aside from the government parties, one must comment on the largest opposition party, the Communists. Communist parties in a few other countries (Japan, for example) have decided to oppose the treaty, and the Italian party itself has been defiant of Moscow on a number of questions, most significantly Czechoslovakia. Yet the Italian party wants to avoid too total a break with Moscow, and thus the NPT will receive a relatively unqualified endorsement.

As in most countries, in Italy also there are some serious arguments in favor of acceptance of the NPT. To begin, signature and ratification would end speculation as to whether Italy will obtain nuclear weapons, precluding anyone from opportunistically starting a weapons ball rolling. If one is convinced that Italy would be worse off having such weapons, it makes sense to preempt the debate before it has even begun.

More pressingly, Italian adherence to the treaty might increase pressures for acceptance of the NPT in other countries. Most Italians probably prefer that Germany never acquire nuclear weapons, a possibility which signature of the NPT helps to avert.

Generally, all acceptances of the treaty make all others more likely; and the chain here can extend to Israel, South Africa, Japan, and Australia. Italy necessarily has some interest in keeping weapons from spreading into these areas. By signing and ratifying, she focuses more world attention on them and increases the likelihood that they will also have to sign.

By signing and coming to terms with the IAEA, Italy assures herself of continued access to fuel and technological assistance from the United States and other signatories. The treaty forbids the rendering of such assistance by any signatory except under Vienna safeguards; since the United States signed the NPT and at last has completed its ratification, Italy will be forced to come to terms with Vienna one way or another, and it may as well do so within the treaty. For some time, Italy ranked high in nuclear electrical power production, but it is now being passed. Such an index can be misleading, however, for one must draw a distinction between installed power capacity and indigenous technology. Italy has all along been heavily dependent on externally-supplied technology, so that it would be quite premature to contemplate going it alone.[2] Racing ahead in installed power production has thus simply made Italy more, rather than less, dependent on United States assistance.

TRENDS IN ATTITUDES

The history of Italian attitudes on a nonproliferation treaty is quite revealing; the NPT illustrates how difficult it is for Italy and other small powers to seriously enunciate positions on arms control. For a long time it seemed doubtful that the superpowers would ever agree on a nonproliferation treaty, since the Russians seemed determined to outlaw the NATO Multilateral Force (MLF) and the United States seemed equally determined to establish that an MLF did not amount to proliferation. During this time, it was easy for the Italian delegation to the ENDC in Geneva to declare

[2] A good account of early Italian progress in nuclear industry can be found in Leonard Beaton, "Capabilities of Non-nuclear Power," *A World of Nuclear Powers*, ed. Alastair Buchan (Englewood Cliffs: Prentice-Hall, 1966), pp. 19-20.

itself wholeheartedly in support of the nonproliferation principle.[3]

A first major reversal of position occurred late in 1967, when Italy suddenly realized that the Soviet Union and the United States were close to agreement.[4] The issue had to be taken seriously in Rome, and a new delegation head in Geneva began to voice the near-nuclear nations' long series of grievances against the proposed NPT. Yet this resistance to the treaty was not prolonged. Italy was governed by a minority Christian Democratic cabinet from June to November 1968, and the government did not want to polarize the country on the NPT question in addition to all the others. In a debate which hardly addressed the issues, Parliament voiced approval of a speedy signature, and it appeared that Italy might sign as early as late August 1968.

After the invasion of Czechoslovakia, however, yet another reversal was in order, as the government announced that it would have to "pause and consider" the appropriateness of signing the treaty. But when the Socialists, the Social Democrats, and the Republicans reentered the cabinet in December 1968, with Pietro Nenni as Foreign Minister, the final decision on signature was determined: Italy signed the treaty at last on 28 January 1969.

Having signed the Nuclear Non-Proliferation Treaty, Italy is likely to ratify. Special circumstances would be necessary to prevent ratification; and only a combination of such circumstances would suffice to kill the treaty. If the United States and the Soviet Union had become involved in an endless delay on ratifying the treaty, all other states, Italy included, obviously would have been deterred from ratifying. It would have been difficult for the superpowers to pressure Rome for a ratification they themselves had not delivered. More directly, a flat West German refusal to sign the NPT might have made Rome reluctant to ratify, either in a spirit of cooperation with Bonn or in the expectation that German abstention had made the treaty meaningless in any event.

As the joint Soviet-American ratification of the NPT has been

---

[3] A July 1967 governmental statement by Aldo Moro heartily endorsing the NPT is reprinted in Stephen D. Kertesz, ed., *Nuclear Non-Proliferation* (South Bend: University of Notre Dame Press, 1967), pp. 99-100.

[4] For less enthusiastic support, see statements reprinted in United States Arms Control and Disarmament Agency, *Documents on Disarmament, 1967* (Washington, D.C.: U.S. Government Printing Office, 1969), pp. 312-19, 527-79, and in *Documents on Disarmament, 1968*, pp. 88-92, 314-16, 714-18, 773-76.

accompanied by West German signature, Italian ratification now would seem to depend on the progress of negotiations between Euratom and the IAEA. Italy is bound by the Euratom treaty not to ratify another nuclear treaty if it in any way conflicts with the Euratom treaty. This might require Rome to stay aloof from the NPT until the Commission or the High Court of the European Communities declares that all contradictions have been resolved, which of course might never happen if a Brussels-Vienna deadlock emerged.

If Euratom's bargaining position depended on the solidarity of a united front, however, there has been reason to doubt Italian resolve even here in resisting the IAEA and the NPT. Since 1968 Rome has devoted considerable effort to seeking a permanent seat on the IAEA Board of Governors. In Italy, as in other countries, it has thus become clear that the fight will not be over once the NPT goes into effect; rather some unresolved problems will remain, and the Vienna Agency may yet be molded into a forum in which the near-nuclear states can present their viewpoints.

In principle, one could hedge one's bets, fighting for increased leverage within an organization while also working to minimize that organization's influence over some substantive area. In practice, however, it is hard to maintain bureaucratic enthusiasm for two such tasks. It may be that Italian proposals for representation on the IAEA Board of Governors reflected little more than a foreign ministry's desire to win as many such posts and points as possible; yet everyone is aware that the Board is much more powerful within the IAEA than such boards usually are in international organizations. The effort to win a permanent seat thus rather suggested throughout the Foreign Ministry that a ratification of the NPT was coming sooner or later and that Italy might as well start living with it. It simultaneously served as a signal to Italy's principal ally (Germany) that Rome could not be counted upon to fight about the Euratom-IAEA issue.

Much of course still depends on the progress of the treaty outside the Euratom area. Invidious comparison can cause some troubles here, inside and outside the "Mediterranean focus." As long as only a few of the Mediterranean states sign and ratify the treaty, some will argue that Italy has no need to be among the first; however, more than this is possible. If Israel confirms the recurring rumors of a nuclear-weapons program, much will change in many of the countries contemplating approval of the NPT, Italy included. Apart from this, new rumors of deployments of Chinese weapons to Albania could fire up the argument that Italy has a special situational need for the nuclear-weapons option that would otherwise be surrendered. Far short of this, several nations may explore the possibility of living with the treaty without joining it, that is, accepting IAEA inspectors on an ad hoc basis wherever fuel is imported from a signatory but retaining the option of skipping inspection whenever indigenous fuel might become available. India and Brazil for the moment seem intent on exploring this option, and if it proves worthwhile, Italians may suggest that their country has everything to gain and nothing to lose by behaving similarly.

Much also depends on developments in Italian domestic politics. The political stability of Italy has been uncertain and remains so. Legal governments range plausibly from center-left coalitions to center-right. Speculation about nonlegal governments coming to power in ways comparable to the army seizure of power in Greece lamentably has also not been totally absent. For the moment such primarily domestic questions will interest Italians far more than nuclear-weapons options. Yet the impact of the domestic settlement on foreign policy questions such as the NPT must be considered.

The most likely course of action would be for Italy to deliver its ratification of the treaty once it has the approval of the Commission of the European Communities in Brussels. A far-right regime, or one that had depended on extra-legal means to acquire power, might however denounce the treaty and rule out ratification indefinitely, paralleling the Spanish and Portuguese examples. Conversely, a government which shifted far to the left could contemplate accelerating ratification in order to show its approval of peace and disarmament, perhaps thereby to strengthen its domestic appeal at the margin by "getting things done." A last disturbing possibility is that political and parliamentary deadlocks in Rome would delay an NPT ratification for reasons which had nothing to do with the treaty itself or with the attitudes of most Italian political factions toward it. Procedural deadlocks are sometimes sustained in struggles for power even at the expense of substantive measures that all parties favor. Whether this would cause great concern abroad as to Italy's intentions in the nuclear field remains to be seen.

12

Paris, Pretoria, Peking

On a superficial first impression, there is a glaring weakness in the efforts of the United States and the Soviet Union to stop the spread of nuclear weapons. The Nuclear Non-Proliferation Treaty limits possession of nuclear weapons to the five nations already in the club—the United States, the Soviet Union, Great Britain, France, and China. Yet only three of the five intend to sign and ratify the treaty. These three will indeed refuse to give away nuclear weapons and will assist peaceful nuclear projects only where there are IAEA safeguards to ensure that no weapons are produced. But will the treaty not be meaningless if France and China are free to offer bombs to anyone that asks for them? Supporters of the NPT can respond that the gap will be filled if most potential recipients of nuclear weapons sign the treaty, thereby promising not to accept such gifts; yet abstentions by nuclear-weapons states may still make a great difference.

Some near-nuclear states are already capable of making bombs without any assistance; however, most such states could do it much more easily with material or technical help from the outside. Indeed, if one were prepared to give complete bombs away, any potential recipient would thereby become a near-nuclear power. France and China may give bombs away, if they want; they have not legally promised not to. They may also give away technical information on how to produce bombs. Other states are similarly significant in that they can sell or give away uranium; what if they also refuse to sign and ratify the NPT?

PARIS

For the moment the attitude of Paris seems reasonably clear.[1] France, its U.N. delegate has said, will not sign the NPT but will behave just as a signatory would.[2] We will of course never know exactly how much assistance France rendered to Israel or whether France left any bombs with its one-time partner.[3] Nor will we know how much truth there is in the rumors that Israeli scientists, in turn, contributed to French progress on the H-bomb. Any such cooperation presumably has been terminated now by political developments other than the NPT.

We can also not be certain that nonsignatory France will interpret the NPT quite as strictly as will signatory states. France will not give away bombs, but would she sell India or someone else crucial components, enabling that country to escape IAEA safeguards and then to make its own weapons? Even the signatories will argue among themselves about which sales are allowed and which are forbidden to states rejecting safeguards. Whenever France and India negotiate any agreement on heavy water, for example, part of the world will again fear an undermining of the NPT.[4]

France plays another important role in nuclear proliferation, apart from having bombs and knowing how to produce them. France has been a member of Euratom and the other European Communities, along with the Benelux countries and Italy and West Germany, two states which have appeared quite reluctant to accept the NPT. Since many of the issues slowing German signature and ratification have involved the reconciliation of Euratom and IAEA safeguards procedures, France might thus be

---

[1] A comprehensive account of the French nuclear program can be found in Lawrence Scheinman, *Atomic Energy Policy in France under the Fourth Republic* (Princeton: Princeton University Press, 1965).

[2] United States Arms Control and Disarmament Agency, *Documents on Disarmament, 1968* (Washington, D.C.: U.S. Government Printing Office, 1969), p. 431 (hereinafter cited as *Documents on Disarmament, 19–*).

[3] See Leonard Beaton, *Must the Bomb Spread?* (Hammondsworth, Middlesex: Penguin, 1966). pp. 77-79, for a fuller account of French-Israeli cooperation on nuclear matters.

[4] See K. Subrahmanyam, "The Path to Nuclear Capability," *Institute for Defence Studies and Analyses Journal* III (July 1970), p. 94.

expected to render some important opinions here, even if it were not already Germany's neighbor, current ally, and former enemy.

At various stages Paris has severely criticized the NPT in the abstract and has announced that it would not sign; yet French governments have been extremely reluctant to voice an opinion as to whether Germany should sign or reject the NPT. It thus seems plausible that France would welcome another definitive German renunciation of nuclear weapons, but would like to avoid pressuring Bonn to make such a move. If Bonn signs and ratifies the treaty, nothing will have been lost. If the United States, Great Britain, and the Soviet Union alienate German public opinion, France will have gained thereby.

But this first, slightly Machiavellian interpretation of French policy does not fully explain French stands within the decision-making processes of the European Communities.[5] A concrete question emerged in 1968 as to when the six governments (or just the five to be inspected) were to begin discussion among themselves on their negotiation position with the Vienna Agency. For a time Paris opposed having any such formal consultations before West Germany had signed the treaty. This could be interpreted as focusing pressure on Bonn to sign, or instead, as seeking to undermine support for the treaty in general, if delay would make satisfactory negotiations less likely. The Netherlands government, anxious to get Bonn to sign the treaty, urged an earlier commencement of negotiations, even before West German signature. At length, France changed its position and agreed. French representatives in Brussels generally have also not been forthcoming with thoughtful or educated opinions on how Euratom and IAEA procedures can be reconciled. This again may simply have reflected a discretely noncommittal French attitude on a treaty Paris will not sign; alternatively, it may have reflected the hope that talks between Brussels and Vienna would reach an impasse. Thus, France, having attacked the NPT, may quite sincerely and straightforwardly have hoped to see it fail to win acceptance in the rest of Western Europe, and perhaps throughout the world.

If Paris has dragged its feet on preparations which might

---

[5] The French interaction with Euratom is fully described in Scheinman, *Atomic Energy Policy in France*, and in Robert Gilpin, *France in the Age of the Scientific State* (Princeton: Princeton University Press, 1968), pp. 403-13.

facilitate a compromise between Euratom and the IAEA, however, a third, even more Machiavellian interpretation remains to be considered. France might indeed ideally still have wanted West German acceptance of the NPT, but only after a showdown between Brussels and Vienna, with a total victory for the IAEA. German acceptance of the NPT in the end, conceding Euratom's total redundance, might thereby have terminated Euratom and its controls over nuclear materials within France.

The relationship between French military nuclear production and Euratom's activities is complicated. Euratom has had access to information about all natural uranium, enriched uranium, thorium, and plutonium entering or circulating within the six member nations. This bookkeeping control terminated only when the materials were clearly consigned for military purposes or moved inside a military facility. The French government presumably has been reluctant for a body such as the IAEA to acquire data on the amounts of uranium which had passed from the civilian sector to the military sector, if only for fear that someone could then begin to estimate the size of the French arsenal. If the IAEA were given access to Euratom's books, Paris might thus have had to insist that all French data, civilian as well as protomilitary, be segregated from the data for Benelux, Germany, and Italy. For all of these reasons, France may have wished to see Brussels adopt a very tough line of resistance to IAEA demands.

If IAEA concessions made possible a compromise that did not threaten French military secrets, however, Paris might still have preferred a less compromising alternative which in the end committed Bonn but killed off Euratom and, with it, all outside safeguards whatsoever in France. France, after all, is the only nuclear-weapons state which has tolerated, as a matter of legal obligation, external controls on its civilian nuclear activities. The United States and Britain have voluntarily submitted some of their facilities to IAEA safeguards, but the Soviet Union has not. While the Anglo-American gesture might not have been very meaningful in any event, France could match it at some later point if Euratom were terminated, but from the legal standpoint of noblesse oblige rather than of a treaty obligation. The NPT does not require such submissions from the nuclear-weapons states.

Negotiations between the Euratom states and the IAEA were thus delayed while the five that would come under IAEA inspection wrangled with France. The West German position was that all civilian nuclear activity in France should come under an Euratom inspection system fully equivalent to the IAEA system under which the other Euratom members would come. Otherwise it was possible that the less strict inspection within France would tempt individual manufacturers of reactors to locate just west of the Rhine rather than just east. When the French government did not agree to this, deadlock and delay persisted until November 1971. The final compromise, which at last allowed the Euratom countries to begin talks with Vienna, reflected a substantial concession to Paris, in that any French plant which engaged in military as well as civilian nuclear activities would be exempt from inspection.

In recent years, France has been very slow to endorse increased appropriations for Euratom inspection activities. Since Euratom has so little else underway, this demonstrates an indifference to the health of the organization and a general resentment of inspection inside France. Frenchman are always present in the inspectorates that monitor West German and other nuclear facilities. One could argue that this should be much more reassuring to Paris than inspectorates of Pakistanis or Argentinians, which the IAEA might someday despatch to these facilities. Thus if IAEA prerogatives threatened Euratom's existence, France should be anxious to maintain Euratom because of the more reliable check on West German intentions that it offers. Yet this very much exaggerates French fears of West German duplicity. If Bonn ratifies the NPT, most French observers will not worry about clandestine bomb projects in West Germany. Removing any strands of legal legitimacy from West German possession of nuclear weapons will be a major accomplishment; Euratom inspection of West Germany does not necessarily offer advantages to outweigh the indignities of Euratom inspection of France.

West Germany, because of its dependence on enriched uranium from the United States, is certainly under increased pressure to accept the NPT. This confirms the predictions of French officials who had urged Germany and other Euratom countries to cooperate in developing natural-uranium-fueled reactors. The pooling of expertise and of materials here over time might have made Europe generally more independent of American influence. It might also have generated financial savings in French moves toward A- and H-bombs. If there was ever a temptation for France to cut its costs by inviting German cooperation on weapons matters, however, today it is in the past. If West Germany's need of enriched uranium today has forced Bonn to accept the NPT and renounce nuclear weapons, France will probably be glad.

PRETORIA

Nuclear-weapons states and/or nations controlling sources of uranium can help would-be nuclear-weapons states to evade the NPT. The three most significant sources of uranium, in descending order, are the United States, Canada, and South Africa.[6] The position of the United States, as one of the nuclear-weapons states that wrote the NPT, is clear. Canada's position, also reasonably clear, is a combination of a moral objection to weapons spread and a general suspicion of the vagaries of the uranium market.

South Africa's position may be more of a problem. Pretoria has not indicated any clear intention of signing the NPT.[7] Also, Pretoria has a policy of not disclosing information on individual sales of uranium. A very plausible commercial argument can be presented in defense of the latter. There is no fluid market or "correct price" for uranium, since governments control the bids wherever it is bought and sold; South Africa would thus be giving its commercial adversaries a needless advantage if it disclosed what price it had gotten, for what amount, in previous sales. Yet the political result is that clandestine sales of uranium to Israel or to Japan become plausible, so that rumors of such sales cannot be rejected outright. Such rumors are not widely circulated within South Africa itself, but this may prove the discipline of Pretoria's civil service more than any government innocence. A country which so successfully avoids disclosing its trade with the rest of Africa and with the Communist world is also capable of concealing uranium transactions.

One can be too quick in accusing Pretoria of giving bombs

---

[6] Uranium sources are clearly catalogued in C. F. Barnaby, ed., *Preventing the Spread of Nuclear Weapons* (London: Souvenir Press, 1969).

[7] *Documents on Disarmament, 1968*, pp. 378-83.

away. Of states immediately interested in nuclear weapons, a fair number would indeed seem to be strongly disqualified from partnership with South Africa. India, Pakistan, and Egypt all have explicitly denounced the white regime. *In extremis*, one of these states might quietly accept uranium even from the devil, but it is difficult to imagine what South Africa could be offered in return. Cash payments might ease South African qualms about sales to Germany, Brazil, Australia, or Argentina, if these states remain outside the NPT system; but cash considerations could also make South Africa support the NPT and the IAEA.

In a time when much of the world has been embargoing South Africa, Japan has been a friend indeed, substantially expanding its trade and investment. If Tokyo were to adopt a nationalistic policy which included the manufacture of its own nuclear weapons in the late 1970s, we might be less certain that Pretoria would withhold the crucial uranium. There has been more speculation on cooperation between Israel and South Africa, some of it maliciously originating from Eastern European sources which like to lump together the regimes they oppose.[8] Israeli involvement in Black Africa may superficially stand in the way of cooperation, as might the occasional anti-Semitism in South Africa, but both could indeed overlook these obstacles if grander projects were involved. One could obviously speculate on a deal whereby both states would obtain nuclear weapons, combining South African uranium and Israeli expertise. Sharing of other military hardware also might bring the two states together, as they are under slightly different kinds of arms embargoes.

Yet there will still be important arguments against South Africa so brazenly defying the world consensus on proliferation. To begin, South Africans will have to be concerned about the revenue they earn from selling uranium to the legitimate peaceful power reactor market. This market now promises to expand rapidly in the 1970s, but the appearance of a sixth nuclear power could substantially hamper this, if the United States and other technologically advanced states became reluctant to license or sell reactors or to process fuel. South Africa has thus several times

[8] See V. Popov, "Nuclear Weapons and Security," from *Krasnaia zvezda*, May 31, 1968, cited in Roman Kolkowicz et al., *The Soviet Union and Arms Control: A Superpower Dilemma* (Baltimore: The Johns Hopkins Press, 1970), pp. 70-115, for implications of Israeli-South African cooperation.

stated that it will not allow its uranium sales to be used to increase the number of nuclear-weapons states.[9]

There is something dignifying and stabilizing for South Africa in being a supplier of uranium, just as there is in being a source of gold. But for this, South Africa's style might be more parochial or xenophobic. South African participation in the IAEA has all along been serious, responsible, and expert; the nation is quite important not only because of its uranium deposits but also because of the advanced status of its nuclear industry. Agency deliberations indeed have been remarkably free of the anti-South African polemics which tend to characterize other specialized agencies of the United Nations.

If the NPT induces the world to tolerate a great expansion of nuclear power production, it would be short-sighted and foolish for South Africa to rock the boat. Of course, the NPT might hurt the uranium market by fracturing it into signatory and nonsignatory blocs with incompatible control systems. If it wins widespread acceptance, however, it can ease barriers to fuel transfers and sales, as the world standardizes on IAEA safeguards to replace bilateral or Euratom arrangements. South Africa's current reluctance to sign the treaty thus might simply reflect a preliminary uncertainty about how the treaty will go over elsewhere, about whether it will stabilize or upset peaceful nuclear development.

For the moment, most of South Africa's uranium sales depend on the availability of some other state's enrichment services, which almost automatically bring safeguards into play. The market for uranium is essentially divided into two clusters of power reactors: the majority, which require enriched uranium as fuel, and the remainder, which utilize natural uranium. A state with reactors requiring natural uranium as fuel can of course directly come to terms with South Africa, and could presumably produce plutonium for atomic bombs directly from South African uranium. States requiring enriched uranium, however, have to find some means to preprocess South African uranium. For the moment there are few such enrichment facilities available outside the current nuclear-weapons states—indeed outside the United States. The Soviet Union, France, and China naturally have facilities for enriching uranium, and Great Britain has a facility which was shut

[9] *Documents on Disarmament, 1968*, p. 379.

down for a time but may now be reopened if the commercial demand for enriched uranium makes it profitable. Proposals for gas centrifuge processes are under study by an Anglo-Dutch-German consortium and in Japan. If these states sign the treaty, however, their facilities and any materials passed through them will be under IAEA safeguards, whether or not South Africa signs.

Concern about South African sales policies might thus be relevant only when natural-uranium-fueled reactors are involved (as in Israel) or when some new and uncommitted means of enriching uranium materializes. Apart from the European and Japanese centrifuge ventures, significant world attention has thus been attracted to Pretoria's claims in 1970 that South Africa was developing an entirely new enrichment process. If and when such processes become commercially viable, the Republic will be able to supply not only its own needs for enriched uranium but also the needs of other states. At that point, Pretoria might find it advantageous to endorse a much more explicit policy, one which couples safeguards to whatever it sells.

As noted above, the Pretoria regime has said that it would not allow South African uranium to be used to expand the number of nuclear-weapons nations. The statement may seem excessively clever, for it apparently allows for sales to any nation that has produced its first bomb indigenously. If countries like Israel or Japan can thus scrape together the raw materials for one bomb, Pretoria could with a straight face sell them the ingredients for any bombs thereafter; like the Israeli statement that Israel would not be the first to introduce nuclear weapons into the Middle East, the South African statement might thus have been composed to seem to be saying more than it really is, soothing the outside world without giving it a grievance to cite later.

Yet all this is not in the South African style. The statement may simply have been sloppily written; or if carefully composed, it may really have been intended to allow military uranium sales to already-nuclear powers, such as France (or Great Britain or the United States). Perhaps Pretoria will thus be dissuaded from delivering natural or enriched uranium to the sixth nuclear-weapons state. Yet one must also consider the Republic's option of itself becoming the sixth. Under present circumstances, by manufacturing nuclear weapons itself, South Africa seemingly would stand to gain less than it would lose. Its conventional superiority over any political opponents in Africa is so clear that it would hardly seem advisable to change the rules of the game. The Republic generally seeks to avoid publicity, and a nuclear-weapons program surely would focus world attention on it all the more. If India, for example, broke the ice by becoming the sixth nuclear power, the Republic could then more easily contemplate conspiring with someone to become the seventh and eighth; for the moment, however, a move toward weapons could rouse great opposition and emotion, within Africa and without.

Yet this is not a definitive argument against all South African interest in the bomb. If great-power intervention is needed to allow the Black African states to break into the White Redoubt, such intervention is not totally unthinkable; resolutions of the U.N. General Assembly supply a mantle of legitimacy. What if the Soviet Union were to assemble an amphibious armada off the coast of South West Africa, merely to implement the rulings of the World Court? Would Pretoria then not at last have a plausible target for any nuclear weapons it had acquired? More correctly, the knowledge that South Africa had the bomb would probably keep the armada project from ever winning such serious consideration in the first place.

It might thus seem entirely prudential for a South African government to begin to assemble the necessary expertise and materials for nuclear-weapons production, even if it is only to be stockpiled for the future, even if some of such expertise had to be purchased by uranium deliveries. Maneuvers of South African armed forces occasionally are conducted on the assumption that the enemy has atomic weapons, which is only reasonable in the 1970s. If no maneuvers to date have included the assumption that the Republic is similarly equipped, it can easily be introduced. How far to go in this direction, and how openly to brandish whatever weapons emerge, will of course depend on the Pretoria regime's own expectations of the likely problems of its foreign policy. If trouble is not already expected, nuclear-weapons activity might still only produce unnecessary trouble.

In at least one respect, Pretoria is rebutting fears that it soon will be seeking after nuclear warheads. The shortest path to such weapons would involve a natural-uranium-fueled reactor free of any American services, allowing the direct manufacture of plutonium. Yet the first South African nuclear power reactors will be fueled with enriched uranium, and the Republic's Atomic Energy Board has published findings that such reactors are definitely more cost-effective than the natural-uranium type. In the long run this may serve to justify a South African investment in uranium enrichment; but plutonium produced in the first power reactors will be subject to agreement with the United States and to inspection by the IAEA, whether or not Pretoria signs the NPT.

If Pretoria wishes neither to obtain nuclear weapons nor to help other "nth" nations to obtain them, it might thus in time become appropriate to sign the NPT. Yet the Republic may achieve its ends more readily without committing itself. South Africa can certainly observe the treaty without signing it. As an important supplier of critical materials rather than a receiver, it can afford to mold a policy of its own, demanding safeguards over most or all of its sales. For the interim, there are some clear bargaining arguments for such a noncommittal position, whereby South Africa would cooperate with the NPT system but would not be bound to do so forever. On a yearly basis, the threat of unsafeguarded uranium sales could thus be held in reserve to deter overly stringent boycotts and embargoes in order to force the United States and Britain to accommodate South African interests where such accommodation is crucial. In this light, it would be optimal for South Africa neither to surrender nor to exercise the option of spreading nuclear weapons.

Aside from this residual threat of selling uranium to weapons producers, the Republic might withhold an NPT signature simply to avoid unnecessary IAEA inspection. There are indeed a few special factors in the South African case that might exacerbate the inspection problem. Because of apartheid, it is always possible that embarrassment might result from the presence of a foreign physicist serving as an inspector; the same dexterity, however, that allows the Republic to treat Malawian diplomats and Japanese businessmen as Europeans could probably be applied to a Nigerian IAEA inspector. The mining industry may pose special problems if IAEA safeguards are to apply to uranium even as it emerges from the ground.[10] (At the moment, IAEA safeguards procedures do not extend to mining, but the NPT text would seem to call for an extension of such inspection to mines as well as to reactors.) Yet

---

[10] Ibid., pp. 381-82.

the mines of Canada and Sweden will be bound by the same precedents of interpretation that apply to South Africa, and these countries have already ratified the treaty, whereas South Africa has not even signed. The Republic's announcement of plans to develop its own new enrichment processes has finally raised real or feigned concerns about commercial espionage.

South Africa's stand on the NPT has not really captured very much attention at home or abroad; the government prefers to say very little about the treaty, and the opposition has not made it an issue. Apartheid and the world's aversion to it dominates what is said and thought about the Republic. Hence there has been little discussion of "loopholes" in the treaty, clandestine bomb projects, or other devious approaches to the nuclear question. South Africans like to think of themselves as a moral people, misunderstood and attacked by the less moral outside world; they profess respect for an international law which they see as misinterpreted and misused against them.

South Africa's statement that it will not contribute to increasing the number of nuclear-weapons nations might be taken abroad as the establishment of a loophole obviously designed to allow Israel or Japan to make a first bomb indigenously and then to make the next fifty with uranium from the Republic. Yet such a loophole is not widely admired or even noticed by the public in South Africa. If the decisionmakers of South Africa were being devious on the nuclear proliferation question, they would thus be pursuing a policy at variance with what most of their constituents regard as proper international sytle. Yet when a country escapes international attention on an issue of this sort, its government also tends to escape attention domestically.

The situation is thus a little paradoxical, in that the kinds of events which would be required to draw domestic and international attention to Pretoria's policies on proliferation are precisely the kinds of unhappy events the NPT is intended to head off. If India or Israel or anyone else produces bombs within the decade, questions about whether all possible sources of bomb material are accounted for will begin to ask themselves more loudly. If they do not, the world may continue to judge South Africa on apartheid only; and definitive assurances on the use of uranium may still be lacking.

PEKING

Peking, a third country capable of undoing the NPT, normally draws the most attention and speculation. Chinese Communist statements in the past have suggested that it would be desirable for all socialist states to possess nuclear weapons. Other pronouncements have declared that no nation can deny another nation the right to such weapons.[11] Denunciations of the NPT have been forthcoming ever since it became clear that the United States and the Soviet Union might agree on such a pact.[12]

Given Chinese interests in various parts of the globe, it has been quite plausible that Peking might deliver weapons to states requesting them. Leaving aside various national liberation groups, which are not likely to receive nuclear weapons from anyone, two possible recipients have drawn some comment, namely, Pakistan and the U.A.R. There are some similarities between the two. Each has a hostile neighbor known to be capable of early manufacture of atomic bombs. Each has despatched delegations of physicists to China; and whatever the intention, this has encouraged speculation on nuclear-weapons assistance from Peking.

At first glance, outsiders might indeed wonder what would prevent such transfers from occurring. Pakistan will not sign the NPT until India does, which probably means never. Egypt signed the treaty under obvious Russian pressure, but will not ratify unless Israel signs and ratifies. Peking is at least nominally committed to great sympathy for proliferation to such states.

If India were definitely to acquire nuclear weapons, any Pakistani regime might very much want to do the same. In the worst of situations, it might be that nothing else would suffice to deter Indian use of bombs. As long as India has not yet manufactured any nuclear explosives, a Pakistani request for such weapons may seem unlikely, for Rawalpindi has clearly preferred a non-nuclear confrontation with its only adversary. New Delhi

[11] See, for example, *Peking Review* VII, October 30, 1964, pp. 6-7.
[12] "A Nuclear Fraud Jointly Hatched by the United States and the Soviet Union," *Peking Review* XI, June 21, 1968, pp. 17-18.

unfortunately senses two adversaries, however; and one of them—China—has already gone nuclear.

If Israel definitely acquires nuclear weapons, the U.A.R. may also want to acquire them. There have been rumors that Egypt sought a promise from the Soviet Union that it would be provided A-bombs if Israel manufactured its own but that this request was rejected.[13] Thus, the U.A.R. might be expected to turn to Peking for the same promise, since Peking will not be bound by the treaty. Indeed, Egypt might want to acquire nuclear weapons from China even before Israel acquired them, for its hopes for peace might not be as high as those of Pakistan for peace with India.

Yet there have been contrary considerations, for both Pakistan and the U.A.R., which make any immediate transfer of nuclear weapons from China less likely than our most pessimistic picture would have it. The U.A.R. has been heavily dependent on Russian military assistance; and Pakistan has long been seeking such assistance. The Russians are opposed to proliferation, and they are at odds with China. For either Pakistan or Egypt to become so visibly involved with Peking might alienate much more immediately necessary assistance in the conventional weapons field, and the atomic bomb is not a panacea which will quickly replace all this. Neither potential weapons recipient has totally burned its bridges to the United States, but each would surely be doing so if it were to announce an agreement for proliferation from Peking.

The reactions of the United States and the Soviet Union might be to go beyond cutoffs in aid. The United States, for example, might have tried to blockade all Chinese shipments to Egypt if it had been made clear that nuclear weapons were being handed over. World public opinion could well be marshaled behind strong sanctions against any nation which so brazenly attempted to become number six in the nuclear club. Much of course would depend on whether either Israel or India had already mobilized opinion against itself by reaching for the bomb.

If the Chinese were willing to give bombs away, Pakistan and Egypt might be reluctant to accept them; moreover, there is good reason to assume that Peking is not at all anxious to give them away. Since the first detonation of a Chinese bomb in 1964, there has been significant hedging in Chinese statements on nuclear

---

[13] *New York Times*, February 4, 1966, p. 1.

proliferation. For example, pronouncements have been made to the effect that assistance from one nation to another is entirely appropriate on peaceful uses of nuclear energy but that it is best for a nation seeking weapons to develop them itself (perhaps thus to retain its independence and self-respect).[14]

Within the last two years there have been more specific statements accusing the Japanese government of aspiring to manufacture nuclear weapons and declaring that Japan under no circumstances should consider doing so.[15] The pro-Peking Communist Party in India has similarly opposed any Indian move toward nuclear weapons.

Ever since detonating its first bomb, the Chinese Communist regime has declared its commitment to a "no-first-use" policy, whereby it would only use nuclear weapons if some other state had done so first.[16] No other nuclear-weapons state has so clearly circumscribed the cases in which it would use its arsenal. The Chinese statements at times have suggested that nuclear weapons would not be used even if an American army were to invade China, as long as the Americans had not used them first. China's giving such weapons to other states would appear to conflict with this apparent desire to keep all wars non-nuclear.

Chinese statements to date thus do not firmly require Peking to refuse requests for bombs, but they certainly allow for such a refusal. The diplomatic reluctance of Cairo and Rawalpindi to become clearly tied to Peking will also make the Chinese reluctant to make firm offers of nuclear-weapons aid. It would be humiliating, at the least, for Peking to be on record, publicly or even privately, with a firm offer, as long as the recipients did not publicly give some indication that they were seriously interested. There is no evidence that private Pakastani and Egyptian talks with the Chinese have ever included discussion of possible proliferation; there are indications that they have not, that the Chinese have not been inviting discussion of this possibility at all.

That proliferation of nuclear weapons might fit in well with the general Chinese approach to world political unrest—a low-cost

[14] See report of press conference with Foreign Minister Chen Yi in *Peking Review* VIII, October 8, 1965, p. 14.

[15] *Peking Review* XI, April 19, 1968, p. 17.

[16] See report of press conference with Chen Yi in *Peking Review* VIII, October 8, 1965, p. 14.

input with a high return of political confusion—has been stated persuasively.[17] Yet the opposite case can also be persuasive. China has not professed to attach great significance to its own bombs; as mentioned above, they are to be used only after others have used them. Peking has attached primary significance in its propaganda to well-founded guerrilla movements. To pass bombs around would reverse all this in the eyes of the world. In the Middle East, it would divert attention from Al Fatah toward the U.A.R. government in Cairo. In general, it invites comparison with the strategic weapons Moscow could have offered; in techniques of guerrilla warfare Peking might have claimed a special advantage, but in bombs it can not.

If one places five or ten atomic bombs into the hands of the Arab governments, the follow-up influence one has earned may not be extensive. A long and gradual guerrilla campaign with Chinese instructors or Chinese machine guns would promise to have much greater leverage and would be more consistent with the "man over weapons" ideology from which Peking has not yet deviated.

Despite American charges against China, moreover, Peking has not been adventurist. The guerrilla movements it has supported by and large seem to have had a modicum of mass following and a chance of winning. To give bombs to small guerrilla movements would be adventurism, perhaps diverting them from the tried and true paths to victory, perhaps making no difference at all. To give bombs to already-established regimes which might topple overnight, as in Ghana, would similarly be adventurism; a war in which nuclear weapons were used by one or two regimes over which Peking had lost its control can hardly be viewed with equanimity in China. If nuclear weapons now are used in anger anywhere, their use anywhere else will be more likely, and Peking has shown every sign of hoping to reduce this risk to itself.

If there is one bona fide satellite or ally of China in the world, it is Albania, but Peking has said nothing about the appropriateness of Albania obtaining its own nuclear weapons. Rumors spread in 1968 that China intended to deploy its MRBMs to Albania, presumably thereby to reach targets in Russia outside Sinkiang's

---

[17] See Morton Halperin, *China and Nuclear Proliferation* (Chicago: University of Chicago Center for Policy Study, 1966), for a persuasive argument that China will actually wish for nuclear proliferation.

striking range. Yet even these would presumably have been under Chinese control, thus not constituting "proliferation," and no confirmation for such rumors is at hand. If one feared a Yugoslav, Western, or Moscow-oriented takeover of Albania, nuclear weapons might indeed be a valuable deterrent; Chinese border conflicts with the Soviet Union similarly might justify a deployment allowing missile strikes along different azimuths. Yet the extent of Chinese caution is shown by Peking's unwillingness to comment openly on or to confirm any such deployments; unadvertised, such deployments would lack much of their deterrent effect. Peking after all described Khrushchev's missile deployment to Cuba as "adventurism" (while it labeled the withdrawal as "capitulationism").

THE REMAINING THREAT

The absence of French, South African, and Chinese Communist signatures to the NPT does not immediately undo the treaty. In our imaginations we are prone to exaggerate the indifference of such states to a further spread of nuclear weapons. Yet to demonstrate that fears are groundless does not terminate their political significance. As long as Peking does not definitely commit itself to resisting proliferation, it will still be possible for weapons to spread from this source. None of the barriers cited above are necessarily permanent.

For the moment, the proliferation scenarios remain linked to other potential nuclear-weapons programs. Israel and India are close to bomb production, and other decisions must adjust to this. For Peking and for any potential bomb recipient, several contingency strategies must thus be considered. Bombs could be supplied quietly before any rival weapons programs reach fruition, to facilitate some grand preemptive sneak attack. Alternatively, announcements that such bombs were already in place could be withheld until after India or Israel announced theirs, thus to establish an immediate counterdeterrent. There are drawbacks to any such transfer prior to explicit provocation; supplying weapons to a seventh member of the nuclear club will be diplomatically much easier than supplying them to the sixth.

A second approach would be to announce mutual agreement that Chinese weapons will be transferred as soon as, and not before, India or Israel acquires them. Unless accompanied by a clandestine prior deployment, however, this risks an attempt by someone to interfere with or preempt the transfer. Everyone today assumes that Peking, Cairo, and Rawalpindi will be more amenable to a transfer if India or Israel obtains bombs; proclaiming this in advance has the advantage of putting it down clearly as a commitment, but it also inspires talk of China having acquired satellites or fears that the bombs have already been transferred.

But the reasoning does not stabilize at any primary level here. Even if China had no intention of introducing nuclear weapons into the Middle East, it encourages such a development by its failure to commit itself to resisting proliferation. By making it possible to imagine Chinese bombs passing into Arab hands, despite any Soviet aversions to proliferation, Peking increases the Israeli public's support for the government's refusal to submit to the NPT and consequently there is an increased chance that bombs will yet appear in the area.

In discussions of arms control, one must be careful to consider thoughts about physical realities, as well as the realities themselves. Even if physical developments pose no real problem, the fears they bring about can take on a life of their own. It may thus only require the naming of a plausible weapons source on either side of a conflict to stimulate precautionary arguments that snowball both sides of a conflict to seek nuclear weapons. "China" suffices on the left, "South Africa" or "France" a little more remotely on the right.

The disincentives to these sources upsetting the treaty are serious, but they may not certainly always remain so. What is worse, one can not guarantee that the images thus aroused will not stampede some part of the world into making the proliferation threat real. It is worth trying to get treaty signatures from France, China, and South Africa. If these can not be obtained, the problem is not at all hopeless, but it is a little more difficult.

13

The Vienna Agency

The Nuclear Non-Proliferation Treaty specifies the International Atomic Energy Agency as the inspecting agent to insure that states do not divert their peaceful nuclear establishments to nuclear-weapons manufacture. Virtually every country opposed to the NPT has thus come to make some criticism of the IAEA and to incorporate such criticisms in its general argument against the treaty. It is clear that the fears and charges elaborated in these arguments are consciously or unconsciously exaggerated; yet the generic conflicts and problems presented are not necessarily unreal, and no one in Vienna or elsewhere will dismiss them outright.

The IAEA, based in Vienna, came into existence in 1957 and since then has engaged in a mixture of inspection and international technological assistance activities.[1] Until the NPT, the IAEA accepted inspection duties under only three circumstances: (1) when it had itself arranged for the transfer of fuel or equipment, (2) when some technical assistance agreement between two states had specified the IAEA as the inspection agent in place of the donor state, or (3) when a state had unilaterally asked the Agency to apply such safeguards over specified facilities. The NPT will thus add a fourth kind of mandate for IAEA safeguards.

[1] For a basic discussion of the IAEA, see Arnold Kramish, *The Peaceful Atom and Foreign Policy* (New York: Harper and Row, 1963), and Lawrence Scheinman, "Nuclear Safeguards, the Peaceful Atom, and the IAEA," *International Conciliation*, vol. 572 (March 1969).

FEARS OF THE AGENCY

The most general problem is simply one of economic cost. It will require money and time to make operations visible and inspectable to an outside monitor. Some of such costs will inevitably fall to the inspected nations; some will be borne by the Agency itself. Generally, the more reliable the inspection or safeguards system, the more expensive and wasteful it will be. Beyond certain degrees of reliability, the costs to improve inspection become enormous. New equipment might have to be designed to make adequate monitoring possible; or more detailed and more costly bookkeeping procedures might be required. Trained personnel will have to be detailed to accompany IAEA inspectors on their rounds. At the extreme, it might even be necessary to shut down operations in a facility to confirm records of its fissionable material. A potential host nation, even if it is voicing premature or exaggerated fears as part of more general propaganda against the NPT, might thus legitimately fear the imposition of such expenses. Where physically more reliable inspection systems become drastically more costly, reasonable men will differ on what is the ideal trade-off.[2] All will thus be interested in what the IAEA itself thinks appropriate.

Apart from fears about the direct cost-security trade-off, the fear is also expressed that the IAEA inspectorate may become imperiously bureaucratic, demanding greater access even where no increase whatsoever is thereby achieved in safeguards reliability. This can come about because of personal vanity, institutional imperialism, or excessive legalism. In the bilateral inspection arrangements of the past, an inspector could easily acquiesce in a commonsense give-and-take, since it was only the prerogatives of his own nation that he was exercising, which national foreign policy might want to relax. An international inspector might sense much less discretion within his mandate, feeling that the letter of the law on safeguards had to be carried out.

---

[2] A full discussion of possible economic damage to non-weapons states under the NPT can be found in H. A. Keller, H. Bolliger, and P. B. Kalff, "On the Economic Implications of the Proposed Nonproliferation Treaty," *Revue du Droit International* (1968), pp. 44-47.

The most extensive IAEA safeguards system and the most articulated complaints against the agency have been in Japan. Much of the Agency's systems experience is derived from the complex of Japanese installations to which it has had access since 1962. Japanese commentators complain that IAEA representatives are much more legalistic and demanding than American or British inspectors were; yet the alleged Japanese preference for the good old days of American or British inspection must be taken with some skepticism. The growth of Japanese nuclear-reactor establishments has probably inevitably generated frictions. In 1962 small research reactors posed much less of a bomb-production threat and thus did not require much more than a friendly visit under the bilateral arrangements with Great Britain and the United States. Any inspection by outsiders today will probably require the inconveniences that the Japanese find so disturbing, such as the maintenance of records in a language other than Japanese. Given Japanese industrialists' freedom to propagandize against ventures which will impose costs on them, such charges against Vienna remain to be verified.[3]

Of course none of the critics of the IAEA is openly arguing that fissionable material should circulate without control (albeit individual industrialists would not mind being trusted in this costfree way). Rather they are contending that other control systems will be sufficient and that the IAEA will thus be redundant, as well as costly and troublesome. Japanese will stress the merits of national self-inspection systems, while West Germans and others will note the advantages of Euratom. If Vienna is determined to inspect all by itself, it will either duplicate or eliminate already existing control bodies.

In trying to predict how a bureaucracy like the IAEA's will really behave under a new and expanded mandate, a wide range of possibilities must be considered. The Agency could be quite reasonable and exercise its duties in a spirit of technological common sense, or it could become impossibly imperious, interpreting its legal mandate as requiring a maximum of intervention and authority. Perhaps a clue can be found in the likely career patterns of Agency inspectors. If they hope to move on to

---

[3] For Japanese complaints on IAEA nuclear safeguards, see *Atoms in Japan* (Japan Atomic Industrial Forum, Inc., 1968), pp. 3-5.

employment with various private firms or national research organizations, it will hardly do for them to be extremely demanding during their tenure with the Agency. We are all familiar with the tendency, at national levels, of regulatory agencies to become cooperative (even overly cooperative) with the industries regulated; perhaps the same phenomenon would arise in safeguards under the NPT. Watching for violations will be dull work in any event, for there may well never be any. Policemen at least have some irreducible minimum of crime to uncover, which keeps their jobs from becoming too dull. The IAEA staffs will probably thus have to be given more positive research duties to balance inspection and maintain personnel morale, which also may impose a limit to the rigor of the safeguards process itself.

Yet other tendencies may dominate. Long-term careers with the IAEA inspectorate might (beneficially or otherwise) preclude agents from preparing jobs for themselves in industry. A new sense of authority may emerge as with some other bureaucracies around the world, as the "prevention of bombs" becomes a shield for all delays, all vanities, all sorts of arrogance.

Perhaps the near-nuclear states would be reassured by the fact that much of the practical experience in safeguards and inspection will have to come from the United States in the near future. If Americans are at all sympathetic to the commercial considerations of German and Japanese industrialists, technical experts from the U.S. AEC might thus be counted upon to keep Vienna from excessive authoritarianism. The IAEA will also be tempted to draw upon the safeguards experience of Euratom, which to date has a larger backlog in the field. This again should put Vienna into a reasonable mood if and when it descends on West Germany, the Benelux countries, and Italy. IAEA technicians will not like to look technologically foolish; they will be cooperative with those who know how to avoid blunders.

On the basis of past warnings that West Germany is conspiring to manufacture bombs, the Russians might urge the Agency to impose more and more intensive inspections on Bonn's peaceful nuclear industry. But the Soviet Union has not been forthcoming with much experienced advice on the operation of safeguards. Indeed it may not possess any such extensive experience from its cooperation with the United Arab Republic, Eastern Europe, or Communist China! Even within the Soviet Union itself, methods of accounting for fissionable materials have apparently been less centralized than those within the United States. The Soviet Union's consistent opposition to IAEA safeguards activities until 1963 hardly fortified any special Soviet influence on inspection theory within the Agency.[4] The experience with Communist China has indeed made the Soviet Union regret its earlier stands. In recent years it has become a stronger and more doctrinaire advocate of IAEA inspection than the United States; yet the fact that the overwhelming bulk of inspection experience lies with the West must count for something.

Since safeguards technology, like all nuclear technology, cannot long be denied or withheld from the IAEA, however, the Agency soon might be able to demand whatever degree of vigilance it sees to be within its mandate. Rather than the influence of individual American or West German technical experts, restraint may thus require the government-level pressures of both the United States and the Soviet Union. The Russians and Americans wrote this treaty and fitted the IAEA into the slot of inspector. Surely they will call the tune as much as anyone, and they should realize that the treaty has to be sold and resold to the near-nuclear states if it is to work.

The good will of both superpowers is not taken for granted in most of the relevant countries, however, and the mandate of the IAEA would hardly be clear if the Soviet Union insisted on greater thoroughness and the United States were alone in defending more flexible standards.

SOME FEARS DISPELLED

Yet despite all these uncertainties, it is indeed plausible that Vienna will not be the unreasonable monster various states claim to fear, for a commonsense attitude is determined by certain structural aspects of the IAEA itself. Whether or not the great powers realized this in 1967, the Agency will have to maintain other functions besides the policing of nonproliferation and will have to be responsive to states other than the current nuclear-

---

[4] See Kramish, *The Peaceful Atom and Foreign Policy*, p. 185.

weapons states. The Agency will not pursue "airtight inspection" "whatever the cost."

Perhaps one of the most significant ways in which the Agency functions in response to other than the great powers is by having economically less developed states represented on the IAEA Board of Governors. Such states are relatively uninterested in the NPT itself, standing neither to gain nor to lose directly from it. They do not share the view that the spread of nuclear weapons is either imminently or drastically dangerous; nor do they have nuclear industries which will suffer from safeguards practices. Such countries are anxious for the IAEA to continue to be a conduit for the sharing and spread of nuclear technology and for it not to become grossly imbalanced toward police duties because of its mandate under the treaty. The demand for balance may become most visible in the process of deciding on the Agency's budget.

At present the superpowers pay a large share of the IAEA budget. Any increase in the costs of inspection will thus be borne disproportionately by the United States and the Soviet Union. Compared to these nations' defense budgets, increases in inspection costs will be so trivial that there will be little monetary constraint on Soviet-American desires for inspection. Yet countries such as the Central African Republic and Algeria will hardly be content to see a larger fraction of their IAEA contributions (however small these contributions are compared with those of the United States and the Soviet Union) go toward paying inspectors rather than technical experts. One alternative would be to have a separate budget for inspection underwritten entirely by the nuclear-weapons states; yet the United States and the Soviet Union would have to be reluctant to accept this, for the logic that they alone should pay the costs would support the view that they alone will benefit from the treaty.

In addition to the truly non-nuclear underdeveloped states, the near-nuclear states fearing excessive inspection will also have a number of seats on the Board; thus the superpowers will have to take into account the views and voting power of these nations also. In fact, the Board of Governors promises to become a forum of complaint against the treaty and its operations for as long as the Agency's practice of extensive consultation with the Board persists.

Throughout the history of the Agency, mandates have been

developed by near unanimity in the Board of Governors.[5] Since IAEA budgets must be approved by two-thirds of the Board, consensus on the funding of any expanded safeguards system will be all the more necessary. The Board's authority is remarkable for a board of an international organization, perhaps in part because it was originally seen as a largely scientific body and its political significance thus was underrated. Even if the United States and the Soviet Union had the votes to push through an expansion of safeguards substantially diverting the budget from technical assistance, it is unlikely that this would happen if it meant that a significant part of the Board would thus go away disgruntled. The needs of the underdeveloped states and the fears of the near-nuclear states supply an incentive for the Agency to be quite modest in its plans for expanding the inspectorate, and they place a constraint on the thoroughness with which it can accomplish inspections.

Besides needing the approval of the Board of Governors, any enormous expansion of the inspection function will also have to meet the opposition of Agency staffs. If the two superpowers should advocate segregating the IAEA's safeguards budget from its budget for more positive research activities, there will be resistance within the Agency on the ground that this would break the technical staffs apart. It is presently envisaged that technicians will go back and forth between safeguards duties and research duties. This would be difficult to arrange with two separate budgets, especially if safeguards had been funded much more generously than other activities. If, for example, the Agency aspires to have all its personnel spend no more than six months of the year on safeguards duties and the rest on research, the two may be locked in step even with separate budgets. Increased inspection would necessitate corresponding budget augmentations for technological assistance; as long as copious developed-state generosity in the latter category is not assured, there will be a restraint on the intensity of safeguards. It will be in the interest of less developed nations to seek to keep the two kinds of IAEA activity in balance; the Agency has an internal motive for seeking the same balance.

For various reasons the IAEA has not felt the NPT to be crucial to its future; nor has it felt its own character to be crucial to the

[5] See Scheinman, "Nuclear Safeguards, the Peaceful Atom, and the IAEA," pp. 17-25.

success of the treaty. The Agency was not really consulted by the superpowers when they were drafting the treaty. The progress (or lack of it) toward winning signatures and ratifications has seemed to depend on events such as the invasion of Czechoslovakia and the American elections or on the specific political situations in countries like West Germany, India, Israel, or Japan. However often Agency safeguards are cited by such countries as an objection to the treaty, all such charges have to be discounted somewhat.

Some of this was illustrated at the Conference of Non-Nuclear Weapons States held at Geneva in September 1968.[6] Serving in part as a forum for the near-nuclear states to voice their cases against the treaty, the conference addressed the Agency in various ways. One resolution did call for a "simplification" of safeguards procedures. But others called for revised representation on the Board of Governors and for more aid to underdeveloped countries through the IAEA. Some Agency representatives were surprised at the extent to which the IAEA was drawn into these recriminations against the superpowers, but Vienna still was not taken simply as the sheriff for these powers, and indeed it does not want to be taken as such.

It is clear that Vienna desires to see the NPT come into operation and to succeed in its purpose. Realistically, the treaty will give the Agency much more function and importance. Yet the safeguards function will grow whether the treaty takes effect or not, since the United States is turning to the Agency for safeguards even under bilateral agreements. Given the uncertainties about the treaty's future and the unpleasantness connected with selling it, Vienna is not hitching its star to it; the Agency will always be seeking to maintain a diversified set of functions.

In the end Vienna may not be crucial to the NPT, simply because inspection is not as crucial as some commentators initially assumed. As noted, having inspectors on the premises is a useful reminder to a nation of its treaty obligations; at the margin such a reminder will deter a government which would otherwise choose to violate the treaty. Given the very large quantities of fissionable material which will circulate in 1978, however, it is less likely that

---

[6] See *Final Document of the Conference of Non-Nuclear-Weapons States*, U.N. Document A-7277.

any reasonable number of inspectors will be able to detect a violation before it occurs, that is, early enough to marshal worldwide intervention actually to prevent bombs from being assembled. Safeguards may embarrass and deter violations, and for this a limited staff of inspectors will suffice; preempting a violation will be much more difficult, perhaps impossible. Inspection and controls by a foreign official are thus important, but more in terms of politics and psychology than of physics. The presence of an outsider reminds a country that the outside world is indeed committed to resisting proliferation.[7]

The relationship of IAEA safeguards to other inspection or material accounting systems can thus be logically ordered.[8] Vienna will bear the hypothetical task of detecting a diversion to military uses after it has occurred. National self-inspection systems will function concurrently but will have to be far more effective. The national systems will ultimately be forced to this higher standard by various requirements apart from basic arms control, such as the need to practice good bookkeeping in nuclear materials management in order to avoid industrial inefficiency, the need to protect the public against the health hazards of radioactive or otherwise toxic wastes, the need to prevent simple theft or other illegal misappropriations of nuclear materials, or the need to maintain authority on what clearly have to be national choices. In effect, the IAEA can accept a lower standard of rigor for its own safeguards, expecting that its postaudit will hasten the day when each nation will impose on itself a rigorous auditing system. At the least this will require that any decision to violate the NPT will have to be a deliberate national decision; it will be much more difficult for a group of scientists to appropriate materials if they do not have the approval of the country's political leadership.

To some extent, this formulation pulls the rug out from under Euratom partisans who cite that IAEA spot inspection systems have not been able to match the comprehensiveness of Brussels's accounting methods. Comprehensive accuracy is not crucial to an international safeguards system; the better Euratom's bookkeeping, the more it approximates Vienna's standards for an acceptable

---

[7] A similar argument is presented by Lawrence S. Finkelstein, in "New Trends in International Affairs," *World Politics* 18 (October 1965): 117-26.
[8] Allan D. McKnight, *Nuclear Non-Proliferation: IAEA and Euratom* (New York: Carnegie Endowment, 1970).

national system. (Vienna may indeed become more comprehensive than Euratom, perhaps with less of a burden, once it has blanket access to entire systems rather than simply to individual facilities, but this is not the central question.)[9]

What Vienna is to supply is thus not a more thorough inspection but rather an "outsider" or "adversary" inspection. If Euratom is to prove the IAEA redundant here, it must do so on the basis that Brussels already has more credible adversary mechanisms, rather than greater accuracy, but here the logical position becomes muddled. It may indeed be true that Frenchmen can be counted upon to distrust West Germans more than Argentinians or Pakistanis can; perhaps World War II is remembered as a European civil war in much of the world. Yet if Euratom and the European Communities aspire to become more than a congeries of mutual distrust, the adversary relationship logically has to end. If Europe succeeds in unifying itself, it will simply become a country like Japan, subject to the same postaudits and the same distrust.

For the short term, Euratom does indeed impose a control on West Germany (and on Belgium, and so on) which Japan cannot impose on Japan. According to its reputation it does so in an efficient and businesslike way; by its existence, furthermore, it presumably contributes to European unity. It might be a real technological and political loss to have Euratom simply fall by the wayside to be replaced by the IAEA plus five national inspectorates. (France will not be subject to Vienna safeguards even if it signs the treaty.) Some ad hoc compromise between the prerogatives of Brussels and those of Vienna thus had to be negotiated.

Yet even in West Germany, the most vocal defender of the Euratom system, there was skepticism about whether Euratom had enough technological substance to make it worth fighting for. The bulk of the functional apparatus of Europe was to be found in the former European Economic Community (EEC) and the European Coal and Steel Community (ECSC). As the Communities are merged, drawing together two strong pillars and one

---

[9] See Arnold Kramish, "The Watched and the Unwatched," *Adelphi Papers*, vol. 36 (June 1967). Also see Lawrence Scheinman, "Euratom: Nuclear Integration in Europe," *International Conciliation*, vol. 563 (May 1967), for a basic description of Euratom procedures and accomplishments.

weak reed, it is not clear that the new pillar would suffer if the reed were omitted altogether.

As noted, Euratom to date has done little but control materials. Its budgetary future for research purposes is extremely clouded. However much one claims that it enjoys methodological advantages over Vienna, Euratom, like the IAEA, would have had a significantly different task in the 1970s from what it had in the 1960s, as the volume of material to control and/or police will have grown so substantially. To judge the future obstrusiveness or efficacy of either agency on the basis of past performance may be to judge them both too easily. West German physicists working on "black-box" automated detection systems and West German industrialists operating nuclear reactor power plants were not convinced that Euratom was of urgent technological value, at least not as convinced as Foreign Ministry spokesmen in Bonn would have had it.

THE RANGE OF UNRESOLVED CONFLICT

Thus there are constraints on the IAEA which preclude it from simply becoming an imperious agent of the United States and the Soviet Union; the Agency is not going to hire tens of thousands of inspectors. At the same time, none of the complaining near-nuclear states seriously hope to circumscribe the Agency's inspection function to the point that it becomes a trivial formality. If the spectrum of possibilities is thus constricted, it will leave a certain range of as yet unresolved conflict. This conflict can be objectified and operationalized in several areas, the two most salient perhaps being the exact number of inspectors to be hired and the nature of the specific safeguards agreements to be negotiated.

The budget for inspection and the number of inspectors must be fixed at some level, and the superpowers will presumably favor higher levels than will the states to be inspected. The exact number of inspectors to be hired thus will be significant to all concerned parties and will pose as important a policy decision as any the Agency has faced.

A second index of IAEA policy is the inspection agreements to be negotiated with each inspected country. A number of earlier documents are relevant here: the IAEA Statute, the Nuclear Non-Proliferation Treaty itself, the IAEA Inspection Documents as amended several times since 1961, and the Safeguards Agreements already in effect with several countries. The drift of events is illustrated by the fact that the Inspection Documents are already considerably less imperious and demanding than the original IAEA Statute. The Statute requires deposit of excess fissionable materials with the IAEA to prevent stockpiling of such materials; the Inspection Documents make no mention of this "requirement."[10]

The reconciliation of these mandates with the wording of the NPT itself could also have caused some litigation and confusion. The Inspection Documents implicitly assume inspection of specific materials supplied under a bilateral or multilateral agreement; the multilateral agreement of the NPT simply subjects all the nuclear materials of a non-weapons signatory to safeguards, which raises the problem of who will specify the nuclear materials.

At the time of their adoption the Inspection Documents in fact imposed some limit on the Agency's access to various countries, since they state that the number of visits depends on the electrical (that is, plutonium) capacity of the reactor and only the larger reactors are subject to continuous inspection or inspection at will.[11] Yet the incipient inflation of electrical power capacities in fact would soon have given Vienna a blank check again, as most reactors will be for power production rather than research, and above the minimum plutonium potential required for continuous access. Nations seeking some defined limitation of Vienna's prerogatives must thus turn to the specific Safeguards Agreements, which have been negotiated on a country-to-country basis up to the present and which specify the exact procedures to be followed in safeguards inspection and the exact degree of inspector access

[10] The text of the IAEA Statute can be found in U.S. State Department, *American Foreign Policy: Current Documents, 1956* (Washington, D.C.: U.S. Government Printing Office, 1957), pp. 915-33. Safeguards Documents 1961, and as amended in 1965, can be found in United States Arms Control and Disarmament Agency, *Documents on Disarmament, 1961*, pp. 21-33, and *Documents on Disarmament, 1965*, pp. 446-60 (hereinafter cited as *Documents on Disarmament, 19—*).

[11] For a discussion of the early workings of the IAEA safeguards, see Mason Willrich, "Safeguarding Atoms for Peace," *American Journal of International Law* (January 1966), pp. 35-54.

for each facility. The Agency has quietly been working on a more standard and uniform model Safeguards Agreement, which would apply to all newcomers to IAEA safeguards and would replace agreements governing countries already under IAEA safeguards.

The sessions of the IAEA Safeguards Committee in the spring of 1971 went extremely well, strengthening the prediction that common sense and accommodation may win out. Extensive participation by West German technical experts presumably committed not only Bonn but the entire Euratom bloc to accepting the compromises as evolved, once the IAEA Board of Governors has adopted them. The verbal formula used has nations "controlling" their own nuclear industries and the Agency "verifying" the accounts of such national control systems. While this formula can clearly still become a source of litigation if the Agency should ever press its verification rights to the hilt or if individual nations should try as much as possible to constrict and contain such verification, the evidence for the moment is that each side is ready to forgo the exploratory litigation of what may never have to be tested. With the United States and the Soviet Union not pressing Vienna to take any doctrinaire positions here, it has a chance of reassuring West Germans, Japanese, and others of the nonobtrusiveness of IAEA inspection and of the dignity of Euratom and the sovereign national governments.

Even if the real issues on inspection are thereby merely papered over, the superficial verbalisms involved can indeed succeed in their purpose. If suspicions never emerge that someone is actually producing bombs clandestinely, it may never be necessary to test how far the Agency can go in "verifying" without obviously duplicating what the nation itself is to do in the way of "controlling."

Normally the Agency would not go ahead in the face of explicit opposition from either superpower. Yet the treaty has to be interpreted as it comes into effect, and some of such interpretation almost inevitably will be determined now. It might thus seem optimal for Vienna slowly to develop something of a mandate which will reassure the near-nuclear states, always working within the constraint that issues cannot be broached so prematurely as to force the superpowers to object. No one knows exactly how many inspectors will be employed on safeguards in 1980, but if low estimates cannot be settled for all time, at least some high estimates can be preempted. Before the need for final decision arises, many other questions may be settled, including the degree of trust between significant nuclear states, the exact number of states party to the treaty, the degree of great-power disarmament, and so on.

In the interim, the balance of power on the IAEA Board of Governors will also become clearer, with the likelihood that the Board will become an important forum for suggestions and complaints from the near-nuclear states. The Agency, from the Director-General on down, will have to be concerned for its mandate, and that mandate will hardly be exclusively to stop proliferation. To cite just one uncertainty, the attitude of Great Britain has changed as it has continued to sue for entry into Europe and Euratom, causing it to cease echoing American and Soviet arguments over the treaty simply as the third nuclear-weapons state. Nineteen eighty is a long way off and difficult to make predictions about. One prediction that cannot be excluded is that by then the Agency will have become as much an advocate for the near-nuclear states as a sheriff for the nuclear states.

It is true that the superpowers will be dominant in real terms for much of the future and that real material strength counts for a great deal. Yet the superpowers chose the IAEA, an already established legal structure, on the assumption that it had some strength of its own, which might indeed be necessary if a barrier to further proliferation is to be effective. By seeking to exploit the structure of the Agency, however, the Soviet Union and the United States have also chosen to be bound by it, and this may at some point frustrate great-power intentions rather than further them.

The treaty's authors had alternative options of dispensing with uniform international inspection altogether or creating a new tailor-made inspection organization for the NPT. The former approach might have made the treaty look so much like the Kellogg-Briand Pact of 1928 that no state would really have been reassured by a neighbor state's signature. The latter approach might have made it impossible to sell the treaty to any of the states that mattered. If significant states still profess to be totally adverse to the Vienna Agency, this does not prove that they have failed to recognize in it an opportunity to have a significant voice in the implementation of the treaty.

POSSIBLE SOURCES OF TROUBLE

If the structure of the IAEA thus necessitates a "reasonable" approach to safeguards under the NPT, this does not guarantee an absence of troubles. Perhaps the worst burden the Agency might encounter under the treaty lies in the Middle East. For the moment it appears that Israel will not sign the treaty and that the Arab states hence will not ratify it. Should this situation change early in the history of the treaty, the world may feel relieved, but the Agency will have acquired its first really onerous inspection burden. India, and therefore Pakistan, will almost certainly not sign the treaty. In every other instance, there will be no presumption of short-term cheating, and Vienna will be able to exercise the tact and flexibility required to preclude real issues from arising. In the Middle East very thorough safeguards procedures will be necessary if the IAEA is to keep its credibility. If the Israeli facility at Dimona must be watched very closely, however, Vienna may not be legally or politically able to justify any less severe procedures for Japan or Sweden, and a great deal of room for maneuver might thereby be lost.

The Middle East now has a pathological tradition of violated agreements, of competitive games of deceiving the U.N. international observers. It may be crucial to establish some reassurance that nuclear weapons are not being produced in the area. Yet one must also consider the possibility that the inclusion of the Middle East will force Vienna to become so much of a policeman as to poison the atmosphere it is trying to create. It will not do to have *Der Spiegel* reporting that Israel (or the United Arab Republic) has produced some bombs and hidden them away, despite the presence of Agency personnel; but it also will not do to have every facility in the world watched as closely as Dimona would have to be in order to prevent such stories.

The Middle East aside, troubles can arise for the Agency from region to region; common sense and good will can postpone many of them, but perhaps not all. With regard to West Germany, a specific problem may or may not arise about the nationality of inspectors. Bonn could probably reject Russian inspectors as a matter of principle for as long as the Soviet Union does not itself accept inspection; other nations have expressed the identical position in the past. The same principle would not exclude inspectors from Communist countries in Eastern Europe, inspectors that Bonn spokesmen have described as Moscow-directed troublemakers. It is possible that such inspectors might indeed make false accusations or impose unnecessarily strict procedures on West German plants; they might even learn operating methods of value to the nuclear industries of Russia or other states. Even if none of these fears were reasonable in light of the IAEA procedures and experience to date, someone in Bonn might be tempted to try rejecting all Communist inspectors without any explication of legal principles, perhaps tolerating only inspectors from Yugoslavia and Rumania. The Agency might thus face a crisis if there had been no Communist inspectors inside the Federal Republic of Germany for three or four years and Moscow thus chose to protest. Better relations between Bonn and Eastern Europe are increasingly plausible, however; common sense may indeed avert such troubles on both sides. It is obvious that everyone in Vienna hopes for such common sense.

Commercial espionage may create problems with inspectors even from outside the Communist bloc. It is indeed likely that a private American firm—much more so than a Russian one—could exploit some new design which was being tested in West Germany and perhaps then beat the West Germans to a sale in South America. Such espionage has not yet been a problem for the IAEA and will only be possible in a small fraction of the field to be inspected, since the bulk of nuclear technology is now in the public domain and can no longer be labeled as secret. Yet an influx of new personnel to fill out augmented inspector staffs could indeed cause some real as well as some imagined scandals here.

A related problem could arise over inspection of the United States and Great Britain, both of which have voluntarily thrown open their peaceful facilities to IAEA safeguards. There is no reason to suspect clandestine bomb production in such facilities, since ample military facilities exist outside the inspected areas. (But is there any reason to suspect clandestine bomb production in a Swedish facility?) If these facilities were to receive their fair share of safeguards personnel, they might require more than 60 percent of the IAEA inspectorate. It is extremely unlikely that

IAEA personnel will be deployed in this fashion; Vienna will rather use access to Great Britian and the United States primarily for training its personnel, a most valuable opportunity for the Agency. Yet this will make it easier for spokesmen of Japan or other states to point to a double standard, claiming that Vienna is engaged in a deceptive hoax here, in that inspection costs and espionage risks will still not be as great for the nuclear-weapons states.

Such token inspections of the United States and Great Britain serve one real function, which only recently has begun to draw attention. Apart from the normal proliferation whereby additional *nations* acquire atomic bombs, there is at least some danger that, in coming decades, proliferation will occur *within* nations, if sloppy accounting and control systems allow private individuals, criminal syndicates, or subnational political units to steal materials for producing their own bombs. This threat is at least as relevant within the five states which are allowed under the NPT to have nuclear weapons as it is in any other state. If fear of embarrassment on the occasional visits by IAEA inspectors drives the U.S. AEC to introduce and maintain tighter control systems than would otherwise prevail, the entire world may have extracted some important benefits from these token inspections.

As suggested earlier, lawyers inside and outside the IAEA have foreseen various difficulties in bringing together the wordings of the NPT and the IAEA Inspection Documents to define the Agency's legal mandate.[12] There may still be several confusions about the Agency's duties under the NPT. Does it have a ticket to survey the entire landscape to ensure no use of peaceful nuclear activity for military purposes? And when is something so clearly nonpeaceful that it has to be denounced? For the moment, the Agency is explicit in avoiding the "landscape" safeguards mandate.[13] Each nation acceding to the treaty will hand over a list of its nuclear facilities, and Vienna will accept the surveillance of them. But what if some country should sign the treaty and then

---

[12] A much more comprehensive discussion of the treaty's legal ambiguities can be found in Mason Willrich, "The Treaty on Non-Proliferation of Nuclear Weapons: Nuclear Technology and World Politics," *Yale Law Journal* (July 1968), pp. 1447-1519.

[13] For example, in the Treaty of Tlatelolco, establishing a Latin American nuclearfree zone, the IAEA is entrusted with the task of safeguarding peaceful nuclear activities, but not of verifying that no nuclear weapons are to be found on the landscape. See *Documents on Disarmament, 1967*, pp. 69-83.

bald-facedly omit a set of facilities from its declaration? Whether the international objection would have to be raised by the IAEA or from outside the Agency is a question that hopefully will never come up, but such a test of the world's resolve is not beyond a malicious imagination.

Nations also might submit everything for inspection and then brazenly launch projects of only marginal commercial value to prepare the way for more rapid bomb manufacture. A reprocessing plant for plutonium, for example, can be justified as preparing fuel for future fast breeder reactors, but it also produces weapons-grade material usable for bombs. A state wishing to intimidate its neighbors with its newly acquired weapons potential might want to proclaim the bomb's existence rather than hide it. The IAEA might be reduced to giving such nations the publicity they desire for their quasi-weapons-programs, while having nothing clearly illegal to denounce.

The procedures that would ensue if there were an unconfirmed suspicion that the treaty had been violated are not totally predictable. The Agency's inspectorate would presumably have to report its findings to the Director-General and to the Board of Governors, which in turn might file a complaint with the U.N. Security Council and General Assembly. Since the veto will apply in Security Council procedures, action might be stymied there if one of the permanent members chose to oppose sanctions against a treaty violator. Yet this underrates the significance of the action of the Board of Governors, for the endorsement of such a complaint would normally mobilize world opinion against the suspected offender. If pressure against a possible offender is to be avoided, the arena will thus have to be the Board of Governors itself, and the same counting of votes would then occur as on other matters involving the Agency with the NPT.

Some states—India (with Pakistan) and Brazil at least—definitely will not sign the NPT in the near future. Yet such states still plan to accept equipment and fuels from signatory nations, which now will be obligated to require IAEA safeguards over such transfers. Vienna thus will have a mixed mandate—blanket coverage of facilities in signatory states and access only to specified facilities in others. Vienna can adopt various approaches to these situations. It could apply pressure for treaty signatures by dragging its heels on the negotiation of specified facility agreements, perhaps arguing that a full-system approach is inherently so much easier and more efficient that the Agency should not be bothered by the older (present) agreements which make a country a checkerboard of inspectable and noninspectable installations. Alternatively it could avoid any pressures on behalf of the treaty, simply continuing to assume any obligations thrust upon it by the NPT or by bilateral agreement.

Some experts believe that states such as India and Brazil are deluding themselves by expecting to be able to maintain two separate fuel cycles, one subject to inspection and one independent of it. The economic advantages of blending the two systems would be great, and over time the entire national nuclear establishment would become contaminated with IAEA access, since Vienna's authority follows any fuel over which it has jurisdiction. This inkblot approach might thus become an obstacle to efforts to produce explosives in "indigenous" facilities while accepting assistance in purely "peaceful" activities. Yet other experts believe that such a separation could be maintained at only moderate economic sacrifice. In any event the inkblot will be potentially fraught with political problems if it should involve following IAEA-safeguarded materials through plants which also process other materials. Does the inspector from Vienna have the right to comment on or denounce the handling of these materials? Can the materials indeed be told apart?

Some additional contention will emerge in defining which equipment can be sold to nations that do not accept IAEA safeguards. The treaty forbids sales of equipment "especially designed" for the handling of fissionable materials, but the Agency in the end may have to render a judgment on the meaning of this wording. The United States has put forward a "trigger list" of such items, which is quite extensive. Countries like Sweden or West Germany will want considerably fewer limitations on what can be sold to India or Brazil.

PEACEFUL NUCLEAR EXPLOSIVES

As one searches for positive IAEA activities to balance out the "negative" role of inspection, one turns naturally to the subject of peaceful nuclear explosives. The nuclear-weapons states stand morally committed by the treaty to facilitate peaceful nuclear explosions, and the U.S. AEC continues to announce that such explosions can indeed bring great benefits to countries utilizing them. Doubts persist on a number of questions. Are such explosions indeed desirable enough and effective enough for any civilian purposes to be worth conducting? Will the superpowers indeed be willing to provide many such explosions now? If so, will it be done largely on a bilateral basis or under the auspices of some international agency? If so, will the agency be the IAEA? If so, how much real authority will Vienna have?

At the limits of our imagination one could suggest giving Vienna the right to decide whether such explosions will be conducted, that is, the right to require them as well as the right to veto them. This clearly would enhance the positive prestige and image of the Agency, but it seems well beyond what the superpowers would agree to. It would not necessitate giving Vienna ownership of such explosives (thus making the IAEA the sixth nuclear power), but even that would not be so illogical. If we trust Vienna to stop proliferation, we can trust it with nuclear explosives; as it is, in principle the Agency could always conspire with some country to run off a few plutonium bombs. Who could then bring suit?

Giving Vienna a veto (requiring that all peaceful nuclear explosions be conducted under international auspices) may be more likely; and this might have been the result of accepting Swedish proposals for text amendments in the NPT.[14] Yet this would again round out the negative rather than the positive side of the Agency's functions. Vienna will most probably become a broker through which nations might place their requests for nuclear explosives, thus removing a need for direct application to a donor nation where this would be politically awkward. This somewhat resembles Vienna's role now as a conduit for technological assistance but does not greatly enhance the Agency's prestige.

It is now almost certain that any international involvement in peaceful nuclear explosions will come through the IAEA. Yet this was not determined from the outset, and the relevant section of

---

[14] See Resolution of *Final Document of the Conference on Non-Nuclear-Weapons States*, U.N. Document A-7277.

the NPT refers merely to "appropriate international procedures." It is not clear at first glance why there should have been any doubts about the relevance of the Vienna Agency here. The IAEA indeed has no direct practical experience in the field, but its experience in the nuclear area is at least better than that of any other international body. Both the United States and the Soviet Union are now determined that the IAEA be the body described in the treaty. Bureaucratic rivalries between the Agency and the U.N. Secretariat can account for part of the movement for an alternative peaceful nuclear explosives body. Aside from that there is of course the continuing resistance to the NPT; there are states which will adopt any arguments, even contradictory arguments, to delay signature. If peaceful explosives will only be offered to countries adhering to the NPT, then the IAEA, in administering them, would become party to a discriminatory mechanism intended to pressure nations into signing.

There will inevitably be nations which refuse to sign while retaining their membership in the IAEA; these may well object to the Agency becoming more and more bound to the provision of the treaty. Other recalcitrants in turn can object that the treaty does not give the Agency enough authority on peaceful explosives, in effect demanding more IAEA involvement in these ventures as their price for signature.

CONCLUSION

As is the case with all international disputes, posturing and overstatement have occurred on both sides of the issue. The novelty of this dispute over the treaty emerges from the unique alignment of states for and against it rather than from the style of diplomatic behavior. In this arena of propaganda and myth, the IAEA has thus been caricatured by opponents of the treaty and has been taken for granted by the treaty's authors. In reality, the Vienna Agency will play a much more complicated role, altogether appropriate to the complications that must ensue as the treaty goes into effect. Diplomats privy to the issue may indeed already see the IAEA for what it is; over time their national interests may allow them to become more candid in describing it.

It is simply too early to tell whether IAEA safeguards will have to be obtrusive to be politically effective. Some of the unknowns hinge on technology, but many others depend on the political climates of the future. National spokesmen who profess to be certain that IAEA safeguards will have to be burdensome are either distorting what their scientists tell them or they are assuming the worst possible political environments. Defenders of the NPT and the IAEA who claim to foresee no problems at all are conversely being overly sanguine.

An examination of the political nature of the Agency itself at least suggests a stronger impetus for moderation and compromise than we might otherwise have expected. The Agency's Board of Governors indeed promises to be the arena in which differences of opinion on the safeguards question will be thrashed out; the IAEA's staffs in turn have internal incentives to become a source of technological good sense, creatively generating modes of compromise rather than issues of dispute.

14

Some Conclusions

The Nuclear Non-Proliferation Treaty is an unusually stimulating issue for any traditional analyst of international politics, if only because it seemingly presents a "reversal of alliances," a refreshing new confrontation in which we can enjoy sorting out the ranks. In what other case have India, Germany, Brazil, Israel, and Japan been aligned against the United States and the Soviet Union? And where do we put Canada and Sweden, and Rumania, Australia, and Egypt? While the American public has hardly become engrossed in the issue, other publics have, and the emergence of new constellations of "countries like India" versus "the superpowers" has revitalized the national interest argument in places where it had become stale. If many of us had come to see international politics as a largely bureaucratic phenomenon, simply playing out intranational disputes on a larger stage, the NPT has given a timely revival to the notion that it is nations that have conflicts, rather than bureaus.

The NPT demands that all but five of the nations of the world renounce nuclear weapons; the authors of the treaty naturally are among the privileged five. Some invidious response from other states is thus hardly surprising. The Indian government at first wished to be quite cautious in its rejections of the treaty, but Mrs. Gandhi's statements in 1968 that India could not sign the NPT "under present circumstances" were greeted with such vehement enthusiasm that they retroactively were recoded as a definitive rejection of the treaty. Antitreaty sentiment, to a lesser extent, has also been roused in Japan and Brazil; the indignity of being asked to submit to the "first unequal treaty of the twentieth century" can seem quite real to those who feel that their national prestige has been slighted.

It is interesting that commentators in many of these states fully expect all Americans to be enthusiastically in favor of the NPT; after all, from the standpoint of national interest the United States must be strongly attracted to a foreign renunciation of weapons which it is not required to match. The news that most Americans are somewhat indifferent to the NPT is greeted with polite disbelief in New Delhi and Rome. It is thus all too easy to see a new line of political confrontation in the NPT, in which an alliance of near-nuclear states is pitted against the United States, the Soviet Union, and perhaps Great Britain. A third bloc of "underdeveloped" states sits on the sidelines, bored by the question of proliferation and concerned lest the issue divert all the most developed states from channeling aid and assistance in their direction. China and France also play ambiguous roles, and a "diplomatic revolution" is thus upon us.

This picture faces severe challenges, however. First, the issues of the old East-West conflict are hardly so dead that nuclear proliferation outweighs them entirely in the making of foreign policy alignment. Second, the internal impact of the NPT is hardly as clear as suggested above. None of the near-nuclear states can really afford to be monolithically opposed to the treaty, just as the United States cannot afford to be monolithically in favor. The crosscurrents that beset nations here must be discussed in terms of substantive issues, but they must also be discussed in terms of bureaucratic factions. In the end, the NPT may be grist for the mills of the bureaucratic analysts as much as for the commentator on international politics.

There is indeed some truth to the raw "national interest" or "foreign policy" conflict between the superpowers and the near-nuclear states. The experienced lawyers of the Italian or West German Foreign Ministries have always known that one does not concede a national prerogative without some quid pro quo. A long series of issues can thus be cited as stemming from the NPT. But the lists cited always run on much too long because such foreign offices instinctively begin to pad their arguments and because these states have cooperated among themselves by swapping arguments against the treaty, however limited the arguments may truly be in their particular relevance. Thus one sees spokesmen for Bonn borrowing Indian terminology to demand that "vertical proliferation" (new weapons acquisitions by the superpowers) be halted at the same time as horizontal proliferation. Since the West German Foreign Ministry had tended to become extremely nervous at the very thought of Soviet-American disarmament talks because such talks could always have led to a thinning out of American conventional or nuclear forces in Europe, this West German objection to the treaty hardly seemed the most serious. Conversely, Indian scientists could be induced to mention the risks of commercial espionage by IAEA inspectors, a warning first sounded in Bonn.

Professional military men almost as automatically come to discover some military need for the atomic bomb. Nuclear land mines in the Himalayas become relevant to India. Nuclear depth-charges are proposed in Tokyo for the antisubmarine forces of the Japanese Naval Self-Defense Force. In every disarmament process of the past, nations were careful not to surrender a defense option too quickly. Why should a Brazilian or Swedish military officer be untrue to his profession now?

There are real issues arising out of the NPT. National prestige is itself a real issue. But the issue of prestige is peculiar in that it seizes upon and distorts other issues as surrogates and vehicles. It is not foolish for the Italian Foreign Ministry to echo Japanese complaints about the treaty; it is only misleadingly dishonest.

POLITICAL ECONOMICS

One of the more serious arguments advanced against the treaty is that it will impose economic disadvantages on countries required to submit to safeguards. Agitation on this point shows up among electrical power producers in Japan and West Germany and, to a lesser extent, in Italy. Inspection will waste time and prevent profits, so the argument goes, such that the NPT should be rejected; commercial espionage will carry off valuable trade secrets, and the whole future of the country's economic growth will be at stake. Businessmen in similar positions in Sweden are surprisingly less prone to testify that Vienna will wreck their economy; but the Swedish entrepreneur is considerably more under the thumb of state policy than his Japanese counterpart and cannot as easily lobby for his private profits by pretending to defend the national interest.

Lest we debunk such claims of businessmen in near-nuclear states too much, one only has to mention American sales representatives who have claimed (and perhaps hoped) that these forecasts were true. If one can beat West Germany out of a reactor sale in Spain by whispering that the NPT will keep the West Germans from supplying fuel, any energetic American firm would be derelict in its duties to its stockholders if it did not do so.

There is also an underlying issue of legal principle or nationalism here, which will fester in any country contemplating adherence to the NPT. It was one thing to ask businessmen to accept international inspectors over property supplied by a foreign country; those who believe in property cannot deny that sellers have the right to place any conditions they please on how and what they sell. It is quite another to impose international inspection on what a nation has produced and will produce for itself. The NPT does not introduce the international inspector, except to some extent within the Euratom area; rather it introduces an entirely new justifying principle for inspection: concessions by treaty to an international consensus, rather than straightforward property rights. The effectiveness of property rights as a control on nuclear-weapons proliferation might soon have collapsed, as the nuclear field became more and more a competitive buyers' market. Yet signing away one's right to truly *own* nuclear resources may seem a viscerally unpleasant requirement to impose on Japan or any country like Japan.

Some of the impact of business in general may have been just the reverse all these years. If American business (together with the U.S. AEC) had not talked up the prospects of nuclear power so much, the implicit threat of proliferation might have been further over the horizon and thus might have given the architects of the NPT more time. And if West German and Japanese entrepreneurs had not maximized profits, these countries would be more able to defy the United States on signing the treaty. For the moment, almost all reactors producing power in the world depend on cheaper, enriched uranium from the United States, because selfish entrepreneurs put their profits ahead of the national independence that could be achieved by using natural uranium. (India and Israel of course are exceptions to this pattern.)

Apart from such inadvertent impacts on the political process, businessmen in the near-nuclear countries might also have some rational reasons for favoring the NPT. Standardization on IAEA safeguards, in lieu of the present mixture of IAEA, Euratom, and national systems, may make it easier to expand the nuclear electrical industry, since some fuel sales and equipment transfers are now delayed or precluded by the incompatibilities of the various control systems. For a country selling uranium, such as South Africa, this standardization might expand a very profitable market; for a country excluded from "Europe," such as Sweden, by killing off Euratom it might win a market entry that would otherwise have been denied. West German nuclear industrialists criticize IAEA inspection practices as insufficiently automated; this is an argument against accepting the NPT, but it is also a sales pitch to the IAEA to purchase West German automation devices once the NPT indeed goes into effect.

As "technological" a question as the Nuclear Non-Proliferation Treaty inevitably raises another internal conflict in all the states affected: that of scientist versus nonscientist. The scientist quite predictably sees his substantive advice being oversimplified and abused by less intelligent politicians and other laymen; he is in favor of free enquiry and full exploration of nuclear physics, but he also feels a moral responsibility to see that the products of this exploration be constricted to peaceful uses.

Hence the impact of the scientific self-image on the political balance becomes interestingly unpredictable. Japanese physicists normally are left-wing, but they alternate between distrust of foreign inspectors and distrust of their domestic government. West German physicists were entrusted with designing bearable inspection apparatus by use of automation techniques and "black boxes," to prove their government's position that their task would be very difficult. Technological enthusiasm led them to succeed much more than expected, thus to some extent undermining Bonn's case against safeguards but also promising sales to the IAEA of German-produced "black boxes." Sweden has accumulated prestige for its technological proficiency on questions of disarmament to the degree that it is reluctant to walk away from the table even when the NPT text is regarded as inadequate. India has sought prestige in the technology of nuclear physics, which then stands in the way of accepting the NPT.

As more and more physicists come to be employed in nuclear industry, their position shifts toward the businessman's distrust of safeguards, toward an idealistic "free pursuit of knowledge" without which science and national economic progress would be stifled. Until there is such employment, or until the IAEA inspector sets foot even in the university research laboratories, the scientist (as in Italy) leans instead toward peace and worldly "international cooperation" and favors the NPT.

ELECTORAL AND BUREAUCRATIC POLITICS

The businessman seeks to maximize profits; other citizens seek to hold elective or appointive political office, if for no other reason than for the sheer fun of it. The NPT is a confusing item for the profit seeker to handle, but it is no less confusing for the office seekers of the non-nuclear nations.

Depending on how it is made to appear in the popular press, a treaty like the NPT can be used for or against an incumbent political party, with varying results for policy. If the treaty is unpopular, as in Japan, it may quickly become obvious that the government has been saddled with an albatross, and all the opposition parties may then rush to oppose it. The treaty is "left" in the sense that it is against armaments; but the Left in Japan can count on the United States to force the governing right-of-center Liberal Democrats to support the NPT (albeit many Liberal Democrats privately hope the Left's opposition will kill it), so that it makes more electoral sense to bend one's ideology against the treaty.

In Sweden, by comparison, it is clear that the incumbent Social Democrats have developed a vested interest in "working for peace"; any treaty thus makes an incumbent record look more productive, although the team which negotiated at Geneva did not see Swedish amendments adopted and has private misgivings on the treaty. Opposition parties in Sweden resent the way in which disarmament negotiations were originally entered into on a multipartisan basis, only to produce an election issue which in 1968 helped the Social Democrats retain power.

A disarmament treaty, which in some sense is what the NPT is, thus carries a lot of symbolic baggage in any internal debate. Regardless of its direct substance, acceptance of such a treaty tends to reflect badly on a nation's military establishment, questioning both the practical need and the moral justification for such forces. Rejection of such a treaty conversely serves as a rebuke to the disarmers of one's political spectrum, typically, but not always, the political left, and a vote of confidence in the military defenders of the realm.

Thus the anti-weapons factions of a number of parliaments become the natural allies of the United States and the Soviet Union, as long as the NPT is not discussed in any great detail. A product of the Eighteen Nation Disarmament Conference in Geneva, a "first step" toward disarmament, the treaty as portrayed by its sponsors meets an emotional response in Italy, in Sweden, and to some extent in West Germany, simply because it is seen as vindicating those who all along have been opposed to weapons. Opponents of the treaty normally do not dare to oppose the treaty on these terms; in Italy the Christian Democrats, the party of the Pope, cannot easily become a pro-weapons party. And the Social Democrats in Sweden, despite their own disquiet about many aspects of the treaty, could hardly throw away their own considerable investment in peace as their issue. Opponents of the NPT must thus attack on relatively circular and elaborate paths in these contexts, and there may not be enough time or public interest then to block the treaty.

Similar domestic effects emerge in the weapons states. Russians and Americans sat together in smoke-filled rooms, polishing drafts of the NPT and comparing notes on bargaining strategy to be applied to states like West Germany, Rumania, and India. Such an experience might indeed have been justified purely because of its novelty. For Lyndon Johnson, the NPT served somewhat to counter the Vietnam War's impact for the 1968 election campaign, proving that Democratic administrations could still bring back fruits of cooperation with the Communist world. If President Johnson had special reasons for favoring the treaty, President Nixon has special reasons for opposing it—Nixon did in fact express reservations throughout the 1968 election campaign. The treaty was, after all, cosponsored by the Democratic Party and the Soviet Union. In the aftermath of Czechoslovakia, it may only have been fitting to slow down signs of cooperation with the Soviet Union. The new Republican administration was moreover committed to shifting attention away from Asia and back to Europe, suggesting more sympathy for Germany's and other states' forebodings about the treaty.

Not all the competition for public office takes place during elections. Holders of bureaucratic office attempt to get ahead by expanding their bureaus' services, operations, and tables of organizations. Another NPT conflict thus has arisen in advanced political communities between individuals whose political futures are tied to the success of already-launched projects and those who have not yet hitched their wagons to such stars. In India the only real support for signing the NPT came from certain senior civil service personnel who feared that rejection of the treaty might terminate the American or Russian aid necessary to the smooth and successful operation of their bureaus. Younger and more detached Indians were much more willing to test the resolve of the superpowers and to run the risk that India might have to go it alone. Indian scientists wedded to the glamour of atomic energy programs would oppose any international commitment which seemed to take the edge off the nation's accomplishments in this field.

A parallel intra-state conflict arises between delegations in the field and governments at home. Until an agreed treaty was finally presented by the United States and the Soviet Union, the NPT was not taken as seriously by states represented at Geneva as it might have been; the small number of persons in an ENDC delegation in effect determined policy, sometimes with bizarre consequences. The Italian position on nonproliferation swung from strong endorsement to severe criticism, and then back again. The positions of some smaller countries oscillated even more widely, leading IAEA Secretary General Eklund to criticize countries which adopted one position at Geneva and another in Vienna.

Yet even as advanced a state as Sweden showed some differences in its positions in Geneva and in Vienna, illustrating perhaps that one delegation was drawn from the Foreign Ministry and the other from the Ministry of Commerce. It may be that the only meaningful governmental position is the final one, so that positions adopted before the actual tabling of the Soviet-American draft should be disregarded. Yet a number of Swedes felt that the actions of Mrs. Myrdal's delegation had foreclosed any options of criticizing or rejecting the treaty. A ten-man (or two-man) delegation may simply be roleplaying and not consulting its home government, but this roleplaying may commit that government nonetheless.

Bureaucratic tensions even exist at the level of international organizations. NPT critics can thus charge that the IAEA will become imperiously obstructive and demanding, according to the natural tendency of bureaucracies to expand their mandates and "build empires." Yet the IAEA does not have a monopoly on international status, and the NPT indeed induced a contest between Vienna and the U.N. Secretariat about who would have jurisdiction over peaceful nuclear explosions when and if such explosives are made available by one of the nuclear-weapons states. When the NPT did not explicitly assign this task to the IAEA, subtle moves were made in New York and Geneva to draw it away from Vienna.

LAW, LOGIC, AND REALITY

Societies advanced enough to have nuclear industries also have systems of law; they therefore have a basic conflict between lawyer and layman, which also confuses debate on the NPT. The clearest examples of a response dictated by lawyers are the West German arguments against the treaty, some of them impossibly esoteric in form. The lawyer's dominance in Bonn can be explained variously as due to a surplus of such persons recruited by the Foreign Ministry or as a reaction to the "illegalism" of the Third Reich. At any rate, it rigorously demanded legal counter-concessions for every concession to the nonproliferation principle, even when other West Germans thought a gratuitous concession here and there would have been politically more effective.

Yet a similar legalism might conversely be attributed to the treaty's authors' production of progressively tighter drafts, which allowed treaty critics to claim that earlier drafts were better. By insisting on inspection when the Soviet Union's original draft made no mention of it, the United States opened a question which may no longer be crucial to the halting of weapons spread but which generates numerous opportunities for haggling and criticism. The treaty has similarly been visibly tightened from draft to draft with regard to peaceful explosives. A non-lawyer might have argued that a shorter and less elaborate treaty would have generated the necessary mandate against nuclear weapons at much less cost, perhaps omitting inspection, perhaps simply letting the test ban prohibit any peaceful explosives projects.

A related conflict has persisted about the desirable degree of logical or legal abstraction, between officials who think and talk of "nth nuclear powers" and those who instead see only "real countries," such as "Israel" and "Japan." It may be that abstraction comes more naturally to lawyers, academics, and logicians, so that the U.S. Arms Control and Disarmament Agency and some of the early American arms-control writers were unduly influenced by this cast of mind. "Professional" foreign service types, on the other hand, who have the day-to-day task of dealing with people and countries in the "real world," saw greater limitations in the NPT approach. A similar debate apparently goes on within the Soviet Union, between those who are anti-"nth" and those who regard the NPT simply as an anti-West German treaty. The more abstract Soviet view presumably gained in influence between 1965 and 1968, when the Middle East War and the India-Pakistan War made a general fear of proliferation seem well-taken. Yet some Russians still persist in telling one and all that they only really fear Bonn and that other states thus must sign the NPT mainly to preclude giving the West Germans an excuse not to sign.

Perhaps the NPT abstraction of defining five "nuclear-weapons states" (with all the rest of the world becoming "non-weapons states") has indeed too much lumped together the possible "nth" countries. Perhaps the first to acquire weapons will thus now have called the Soviet-American bluff, visibly setting a precedent, encouraging all the rest to defy the NPT mandate in the same way. Yet the costs of abstraction can be depicted too pessimistically here. In a paradoxical way, the NPT may instead carve a unique sixth slot for any nation that can cite special circumstances to opt out of it. If India now refuses to sign and everyone else does sign, we all may start talking about "India's special strategic situation," implicitly resigned to New Delhi joining the nuclear club whenever she gets around to it. As noted, the world may then consist of nuclear-weapons states, weapons-renouncing states, and India. Preventing West German bombs in the end may simply have lowered the barrier to Indian bombs, and the United States and the Soviet Union will have to contemplate the character of the trade-off.

For a time Indian diplomats kept referring to "countries like India," which also had to oppose the treaty. When signs emerged however that the United States and the Soviet Union by default might have to regard India as a special case (because of China, and because India was not on the losing side in World War II), the Indian government wisely decided to cease leading the pack in opposition to the NPT, but rather to revert to stating its own position, which might have little relevance to other countries. If India indeed goes ahead to manufacture nuclear explosives, it will wish to deny that it has set a precedent thereby and that there are indeed any "countries like India."

THE FUTURE

There will thus always be some doubt about whether the NPT was worth its costs and about whether it might have been better handled by the powers proposing it. Has proliferation not actually been encouraged in the internal debates the treaty has caused? Wouldn't a more subtle policy have been more effective? Aren't the assumptions of Soviet-American cooperation premature?

But the costs and the conflict about the NPT have been obvious, while the hypothetically lower costs of the more subtle approach can be easily underestimated. It was perhaps inevitable that the potential for nuclear weapons would produce some unpleasantness in the world. One form of unpleasantness is that bombs get produced and deployed. The alternative form is that constrictions are imposed to prevent this. The endorsement of the treaty by a hundred or more nations will clearly reduce the legitimacy of a nuclear-weapons program in these and other nations. What was once a neutral or attractive idea is now somewhat tainted and suspect. Nuclear weapons are not just the "latest thing" anymore; they lack respectability. One must only remember that Sweden in the late 1950s was seriously considering a bomb program to see how casually a nation might have decided to join the club; France's decision was surely a calm one. Without the NPT, a politician seeking a new electoral issue for the future could have suddenly become a bomb advocate. With the unpleasant debates that have occurred on the NPT, the bomb is certainly not a new issue anymore; perhaps the public involved has thus been inoculated against a desire for nuclear weapons.

One can point to ways in which the superpowers could obviously have handled the treaty better. Earlier agreement on a text would have produced a more favorable response. It may have been a mistake to publish a text without an Article III on inspection, which thus allowed more than usual attention to be diverted to this particular aspect of the NPT system. Invading Czechoslovakia most assuredly hurt the treaty's chances.

Yet to identify mistakes is not always the most relevant exercise in political analysis. Some mistakes may be inevitable, just as it is inevitable that 65 percent of the heat in a fireplace goes up the chimney and only 35 percent warms the room. If all heating systems are wasteful, we still do not elect to freeze. Earlier Soviet-American agreement may have been impossible, given the special needs of each alliance. Publishing an agreed text lacking an Article III may have been necessary in order to instill a sense of momentum, even if it attracted more attention to the inspection issue. And the importance that Russians attached to Czechoslovakia may have been enough to outweigh whatever harm was done to gaining a general acceptance of the NPT.

Whether nuclear weapons will be spread or contained has not yet been decided. The treaty is neither a success nor a failure. Now that the treaty has gone into legal effect, commentators will be tempted to finalize their conclusions and turn to other issues, but many such conclusions must remain premature.

One hypothesis will be that legal acceptance of the NPT is now assured. Yet there is still reason to doubt that Japan will ratify the treaty, and we can be almost certain that India will never sign. Important nations in the Middle East moreover will either not sign or not ratify. Another will be that the treaty will at least function well for those nations which do ratify and accept the NPT system. Yet a basic issue will remain on whether thoroughness or thrift should be stressed in safeguards under the IAEA, and this will be a subject of contention for a decade or more.

A diametrically opposite conclusion, also premature, is that the signatures of the NPT will prove entirely meaningless, in the manner of the Kellogg-Briand Pact, so that nations will merrily move ahead to acquire nuclear weapons. Signatures are indeed signatures, and they will count for something more substantial in a number of internal political debates. Bombs can no longer be acquired as absent-mindedly as before; in many cases they may be forsworn primarily because of commitment to the NPT.

A fourth hypothesis is that the success or failure of the treaty now will depend entirely on keeping the number of nuclear powers at five. The sixth, or assuredly the seventh, membership in the nuclear club presumably would destroy the entire treaty system. This argument is indeed good scare propaganda with which to dissuade Israelis or Indians from reaching for the bomb, but it is too early to tell whether it is entirely true. Even if superpower resolve on India is clearly fading, even if Israel elects to be the first member of the nuclear club to accumulate weapons without ever test-detonating one, we may yet see the NPT used to rally the rest of the world into an abstinence from further proliferation.

Index

THE JOHNS HOPKINS UNIVERSITY PRESS

This book was composed in Baskerville text and display type, printed
on 60-lb., Warren's Sebago, regular finish, paper, and bound in Columbia
Bayside linen cloth by Port City Press, Inc.

Library of Congress Cataloging in Publication Data

Quester, George H.
    The politics of nuclear proliferation.

    Includes bibliographical references.
    1. Atomic weapons and disarmament. 2. Atomic
weapons. I. Title.
JX1974.7.Q47   327′.174   73-8119
ISBN 0-8018-1477-4